Lecture Notes of the Institute for Computer Sciences, Social Informatics and Telecommunications Engineering 80

Peter Coles Sanmay Das
Sébastien Lahaie Boleslaw Szymanski (Eds.)

Auctions, Market Mechanisms, and Their Applications

Second International ICST Conference
AMMA 2011
New York, NY, USA, August 22–23, 2011
Revised Selected Papers

Springer

Volume Editors

Peter Coles
Harvard Business School
Boston, MA 02163, USA
E-mail: pcoles@hbs.edu

Sanmay Das
Rensselaer Polytechnic Institute
Department of Computer Science
Troy, NY 12180-3590, USA
E-mail: sanmay@cs.rpi.edu

Sébastien Lahaie
Yahoo! Research
New York, NY 10018, USA
E-mail: lahaies@yahoo-inc.com

Boleslaw Szymanski
Rensselaer Polytechnic Institute
Department of Computer Science
Troy, NY 12180-3590, USA
E-mail: szymab@rpi.edu

ISSN 1867-8211
e-ISSN 1867-822X
ISBN 978-3-642-30912-0
e-ISBN 978-3-642-30913-7
DOI 10.1007/978-3-642-30913-7

Springer Heidelberg Dordrecht London New York

Library of Congress Control Number: 2012939046

CR Subject Classification (1998): J.1, K.1, C.2.4, C.3, H.2.8, H.4

Typesetting: Camera-ready by author, data conversion by Scientific Publishing Services, Chennai, India

Printed on acid-free paper

Springer is part of Springer Science+Business Media (www.springer.com)

Preface

The second edition of the biennial Conference on Auctions, Market Mechanisms and Their Applications (AMMA) was held in August 2011 in New York City, following on the success of the first edition, held in Boston in 2009.

AMMA is a forum for advances in the theory and practice of auctions and market mechanisms. The goal of the conference is to bring together researchers from computer science, economics, and business schools as well as industry, with common interests in issues that arise in all stages of deploying market mechanisms. In addition to more traditional academic papers, the conference also focuses on experiences from the real world such as case studies and new applications. This year, the technical program featured 22 papers and case studies by computer scientists and economists on topics such as school choice, markets for housing, energy, and advertising, prediction markets, and the theory of market design.

The conference also featured two invited talks on market design in both the public and private sectors. Peter Cramton of the University of Maryland presented "Medicare Auctions: A Case Study of Government Market Design," discussing the design of an auction for medical equipment procurement for Medicare. Jeremy A. Smith of SecondMarket, Inc., presented "Dutch Auction. English Auction. Manhattan Auction?", discussing auctions for esoteric assets.

AMMA offers both archival and non-archival tracks to accommodate the publishing traditions of different disciplines. These proceedings contain the full papers accepted to the archival track, as well as the abstracts of papers presented in the non-archival track. For full versions of non-archival papers, please contact the authors or visit their websites.

October 2011

Peter Coles
Sanmay Das
Sébastien Lahaie
Boleslaw Szymanski

Conference Organization

Steering Committee

Peter Coles	Harvard Business School, USA
Vincent Conitzer	Duke University, USA
Sanmay Das	Rensselaer Polytechnic Institute, USA
Sébastien Lahaie	Yahoo! Research, USA
Michael Ostrovsky	Stanford GSB, USA
David Pennock	Yahoo! Research, USA
Boleslaw Szymanski	Rensselaer Polytechnic Institute, USA

Organizing Committee

General Chairs

Sanmay Das	Rensselaer Polytechnic Institute, USA
Boleslaw Szymanski	Rensselaer Polytechnic Institute, USA

Technical Program Chairs

Peter Coles	Harvard Business School, USA
Sébastien Lahaie	Yahoo! Research, USA

Local Chair

Ted Brown	CUNY, USA

Conference Coordinators

Åza Swedin	European Alliance for Innovation
Elisa Mendini	European Alliance for Innovation
Ruzanna Najaryan	European Alliance for Innovation

Webmasters

Meenal Chhabra	Rensselaer Polytechnic Institute, USA
Yousaf Shah	Rensselaer Polytechnic Institute, USA

Program Committee

Itai Ashlagi	MIT Sloan, USA
Ned Augenblick	Berkeley Haas, USA
Moshe Babaioff	Microsoft Research, USA
Martin Bichler	TU Munich, Germany
Craig Boutilier	University of Toronto, Canada
Eric Budish	Chicago Booth, USA
Ruggiero Cavallo	Yahoo! Research, USA
Yeon-Koo Che	Columbia University, USA
Yiling Chen	Harvard University, USA
Vincent Conitzer	Duke University, USA
Florin Constantin	Georgia Tech, USA
Sanmay Das	Rensselaer Polytechnic Institute, USA
Ben Edelman	Harvard Business School, USA
Clayton Featherstone	Harvard Business School, USA
Jon Feldman	Google Research, USA
Amy Greenwald	Brown University, USA
John Hatfield	Stanford GSB, USA
Patrick Jordan	Yahoo! Research, USA
Ian Kash	Harvard University, USA
Scott Kominers	Harvard University, USA
Kate Larson	University of Waterloo, Canada
John Ledyard	Caltech, USA
Robin Lee	NYU Stern, USA
Kevin Leyton-Brown	University of British Columbia, Canada
Ben Lubin	Boston University, USA
S. Muthukrishnan	Rutgers University, USA
Michael Ostrovsky	Stanford GSB, USA
David Pennock	Yahoo! Research, USA
Maher Said	Olin School of Business, USA
Rahul Sami	University of Michigan, USA
Sven Seuken	Harvard University, USA
Boleslaw Szymanski	Rensselaer Polytechnic Institute, USA
William Walsh	Google Inc., USA
Michael Wellman	University of Michigan, USA
Makoto Yokoo	Kyushu University, Japan

Table of Contents

Market Design and Interfaces

Internet and Information Markets

Combinatorial Mechanisms

Auction Theory

Matching and Double Auctions

Novel Applications

Strategyproofness for "Price Takers" as a Desideratum for Market Design

Eduardo Azevedo and Eric Budish*

Abstract. We distinguish between two ways a mechanism can fail to be strategyproof. A mechanism may have manipulations that *persist with market size* (first-order manipulations); and, a mechanism may have manipulations that *vanish with market size* (second-order manipulations). We say that a non-strategyproof mechanism is *strategyproof in the large* (SP-L) if all of its manipulations vanish with market size; that is, if it is strategyproof for "price takers". We put "price takers" in quotes because our notion is not limited to mechanisms that explicitly use prices. Our main result is that, given a mechanism with Bayes-Nash or complete information Nash equilibria, there exists a prior free mechanism that is SP-L and that coincides exactly with the original mechanism in the limit. It coincides approximately in large finite markets, with exponential rate of convergence. Thus, while strategyproofness often severely limits what kinds of mechanisms are possible, for our class of problems SP-L does not, and hence may be a useful second-best. We illustrate our concepts with examples from single-unit assignment, multi-unit assignment, matching and auctions.

* Harvard University (azevedo@fas.harvard.edu) and University of Chicago Booth School of Business (eric.budish@chicagobooth.edu). For helpful discussions we are grateful to Susan Athey, Aaron Bodoh-Creed, Gabriel Carroll, Jeff Ely, Drew Fudenberg, Jason Hartline, John Hatfield, Richard Holden, Emir Kamenica, Fuhito Kojima, Scott Kominers, Jacob Leshno, Paul Milgrom, Roger Myerson, David Parkes, Parag Pathak, Nicola Persico, Canice Prendergast, Ilya Segal, Lars Stole, Glen Weyl, and especially Al Roth.

P. Coles et al. (Eds.): AMMA 2011, LNICST 80, p. 1, 2012.

Market User Interface Design
(Extended Abstract)*

Sven Seuken[1], David C. Parkes[1], Eric Horvitz[2], Kamal Jain[2], Mary Czerwinski[2], and Desney Tan[2]

[1] School of Engineering & Applied Sciences, Harvard University, Cambridge, MA 02138
{seuken,parkes}@eecs.harvard.edu
[2] Microsoft Research, Redmond, WA 98052
{horvitz,kamalj,marycz,desney}@microsoft.com

1 Introduction: Market Design and UI Design

Electronic markets are becoming more and more pervasive but a remaining challenge is to develop user interfaces (UIs) to promote efficient market outcomes. This can be a challenge in markets with a large number of choices, yet traditional economic models do not consider that humans have cognitive costs, bounded time for decision making, and bounded computational resources.

Behavioral economists have begun to explore the cognitive costs associated with decision making in complex environments [1], but until now, the market design community has largely ignored the intersection of market design and UI design. Gajos et al. [2] have designed a system that can *automatically* generate UIs that are adapted to a person's devices, tasks, and abilities, but not for market domains. In our own previous work [3], we have introduced the goal of designing *simple* and *easy-to-use* interfaces for electronic markets, in particular for domains where users repeatedly make decisions of small individual value.

In this work, we propose a new research agenda on "market user interfaces" and present an experimental study of the market UI design space. A market UI can best be defined via two questions: first, what information is displayed to the user? Second, what choices/how many choices are offered to the user? Our goal is to design market UIs that make the decision-making task easier for the users and lead to more efficient market outcomes. Thus, the research question we want to answer is: *what is the optimal market UI given that users have cognitive costs?*

2 Experiment Design

To answer this question, we conducted a lab experiment with 53 users. We situate this experiment in the domain of 3G bandwidth allocation, assuming a fictitious market game with inter-temporal budget constraints. Figure 1 presents

* Working paper available at:
www.eecs.harvard.edu/~seuken/MarketUIs2011.pdf

P. Coles et al. (Eds.): AMMA 2011, LNICST 80, pp. 2–4, 2012.

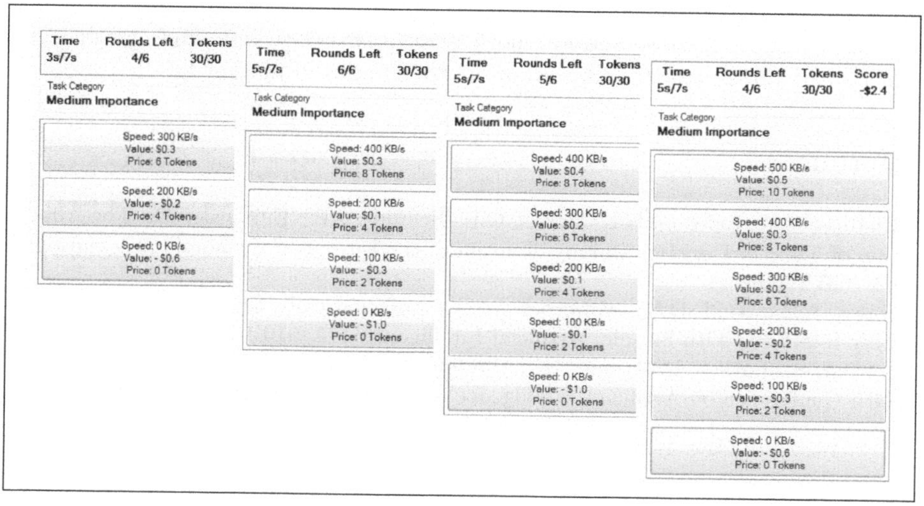

Fig. 1. Screenshots for the game used in the experiments with 3, 4, 5, and 6 choices

four screenshots of the game with 3, 4, 5, and 6 choices. Each game has 6 rounds and a user starts with 30 tokens which he can use to select different connection speeds. Each available speed has a *price*, in terms of tokens, and a *value*, in terms of dollars. Upon selecting a particular speed, the corresponding amount of tokens is deducted from the user's budget, and the corresponding value is added to the user's score. Determining the optimal choice is difficult, because of the budget constraint and because the values and prices change stochastically in each round. Furthermore, the user has to make a decision within a time limit (7 or 12 seconds) before the lowest speed is automatically selected. Each user played 32 games and was paid his total score in the form of an Amazon gift card.

3 Overview of Results

Our experimental results illustrate the importance of taking behavioral factors into account for the design of market UIs. Our findings show that users make more mistakes with a larger number of choices, users select the optimal choice more often when its position is high, and users exhibit strong loss aversion.

When varying the "number of choices," we find that *overall efficiency* significantly increases as we go from 3 to 4 to 5 choices, and then slightly decreases as we go from 5 to 6 choices. When changing whether the composition of the choice sets stays fixed or adaptively changes according to the game situation, we find that adaptive choice sets lead to higher efficiency, with high statistical significance. Finally, we fit a quantal-response model to users' behavior and evaluate an *optimized market UI*. Here, we find the surprising result that the optimization actually *decreases* overall efficiency. Most interestingly, in particular the *more rational users* suffer from the optimization procedure, while no statistically significant effect can be observed for the *less rational users*. This finding naturally suggests our future research direction on "personalized market UIs."

References

1. Chabris, C.F., Laibson, D.I., Morris, C.L., Schuldt, J.P., Taubinsky, D.: The Allocation of Time in Decision-Making. Journal of the European Economic Association 7, 628–637 (2009)
2. Gajos, K.Z., Weld, D.S., Wobbrockc, J.O.: Automatically Generating Personalized User Interfaces with Supple. Artificial Intelligence 174, 910–950 (2010)
3. Seuken, S., Parkes, D.C., Jain, K.: Hidden Market Design. In: Proceedings of the 24th Conference on Artificial Intelligence (AAAI), Atlanta, GA (July 2010)

Evaluating Hidden Market Design

Florian Teschner and Christof Weinhardt

Karlsruhe Institute of Technology (KIT)
{teschner,weinhardt}@kit.edu
http://www.im.uni-karlsruhe.de

Abstract. Electronic markets are increasingly gaining importance in the coordination of complex allocation problems. One reason for market failure is the inherent complexity excluding non-sophisticated users. Recently researchers proposed the idea of hidden market design which merges the fields of market design with user interface design in order to make complex markets accessible to a broader audience. One way to accomplish that is to simplify the market interface. Thus far it remains empirically unclear how using such interfaces affects market efficiency and individual trading performance. In a prediction market for economic variables, traders can choose between a standard trading interface, and one that hides most market complexities. We find that market participants using a simplified trading -*hidden market*- interface are more likely to submit profitable orders.

Keywords: Hidden market design, Prediction markets, Trading interfaces, Marco-economic variables.

1 Introduction

The Internet has increased the number of complex (e.g. Energy, P2P resource sharing) markets dramatically. As more and more non-sophisticated users have to interact with complex markets, the question arises how to provide interfaces for such users to participate. Promoting the idea of *Hidden Market Design*, researchers have recently identified the need to merge interface and market design [17]. The main idea is to hide or reduce market complexities while maintaining economic efficiency. One way to accomplish that is to simplify the market interface. However it remains empirically unclear how simplified trading interfaces effect market efficiency and individual trading decisions.

We study a prediction market called Economic Indicator Exchange[1] (EIX) forecasting economic indicators such as GDP, inflation, investments, export and unemployment figures in Germany. The basic idea of prediction markets is to trade contracts whose payoff depend on the outcome of future events. Market participants form expectations about the outcome of an event (e.g. the economic growth in the next quarter). Comparable to financial instruments, they buy if

[1] www.eix-market.de and eix.handelsblatt.de

P. Coles et al. (Eds.): AMMA 2011, LNICST 80, pp. 5–17, 2012.

they find that prices underestimate the event in question and they sell if they find that prices overestimate the probability of an event.

The advantages of this research setting are twofold. First, from an individual perspective market participants interact in a repeated decision-making environment closely reassembling decision-making in financial markets. Secondly, as the outcome of events in prediction markets is finally known, we can ex-post measure the participants' trading performance. In a field experiment with more than 600 participants and over 40,000 trading decisions, participants can individually choose between two trading interface types. One interface type is a standard trading interface, whereas the other hides most market complexities. Recording through which interface an order is submitted allows us to link trading performance and interface type. Evaluating the hidden market design paradigm from an individual perspective, we find that alternative trading interfaces change participants' behavior. Furthermore, and against naive intuition, we find that orders submitted through a simplified interface are more likely to be profitable compared to orders which are submitted through the default trading interface.

The remainder of this paper is structured as follows: the second section presents a review of related work in the hidden market domain. Additionally a short introduction to prediction markets is given. The third section details the field experiment setting and the framing of the participants' trading process. The subsequent section first presents some descriptive data and then introduces the evaluation methodology. Specifically, we use market measures to separately analyze trading performance and trading behavior. In section five we link the interface types to trading outcome and interpret the results. Finally section six concludes this paper.

2 Related Work

In the following section we will first present related work in the hidden market domain and then introduce related work in the prediction market area.

Hidden Market Design

Challenged by the rise of complex markets (e.g. Energy, P2P resource sharing) in which non-sophisticated users find it hard to interact, Seuken et al. proposed the idea of Hidden Market Design. *'The Hidden Market Design challenge is to find new techniques and approaches towards designing and building hidden markets for non-sophisticated users. The primary goal [..] is to find the right trade-off between hiding or reducing some of the market complexities while maximizing economic efficiency attained in equilibrium.'* [17]. Hence the goal is to lower the entrance barriers (e.g. market complexities) for non-sophisticated users to participate in markets. The simplification can be achieved by either changing the user interface or adapting the market rules. Following the idea they design a market-based P2P backup application [18]. In the paper they address both aspects; the user interface eliciting participants' preferences and the market rules, standardizing the market interaction. However it remains unclear how the simplified trading interface effects market efficiency and individual trading decisions.

Decision Processes in Trading Environments

To our knowledge there exist no empirical work on decision processes in trading environments with focus on the trading interface. Kauffman and Diamond [13] highlight the importance of research on behavioral decision making and information presentation effects. They examine how behavioral effects may become operative in screen-based securities and foreign exchange trading activities, where users can choose among information presentation formats which support trader decision making. They present a model to identify where and how information, heuristics and biases might effect decision making in the trading environment. There exists -to our knowledge- no empirical work linking the decision making in continuous markets to the trading interface. In the domains of decision support systems and online shopping environments the influence of the interface on decision behavior has been repeatedly demonstrated. To summarize previous work the amount and control of information, as well as the information representation [22,23] does influence user behavior. On the one hand information control improves performance by improving the fit between actions and outcomes. On the other hand in terms of cost (disadvantages), information control requires the user to invest processing resources in managing the information amount and flow. As a conclusion information control has both positive and negative effects on performance [1]. The two tasks of processing and managing information are related and codependent. Turning to the optimal pool of available information in decision support systems, empirical work has shown that users can handle only a certain amount of data. Malhotra [15] concludes that individuals cannot optimally handle more than ten information items or attributes simultaneously. Testing decision accuracy Streufert et al. [20] show that as information load increases, decision making first increases, reaches an optimum (information load ten) and then decreases.

Prediction Markets for Economic Derivatives

Prediction markets have a long track of successful application in a wide area ranging from political to sport events sometimes outperforming established forecast methods [3,14,9]. The roots of their predictive power are twofold; the market provides the incentives for traders to truthfully disclose their information and an algorithm to weight opinions [2]. They facilitate and support decision making through aggregating expectations about events [11,4,12]. The most basic trading mechanism for prediction markets is based on a continuous double auction for one stock which represents the outcome of an event. The stock will pay 1 if an event has the predicted outcome and else the stock will be worthless. Market participants form expectations about the outcome of an event. Comparable to financial markets, they buy if they find that prices underestimate the event in question and they sell a stock if they find that prices overestimate the probability of an event.

In an attempt to set up a market to predict economic variables in 2002 Goldman Sachs and Deutsche Bank created the so called 'Economic Derivatives' market. It tries to predict macro-economic outcomes such as ISM Manufacturing, change in Non-Farm Payrolls, Initial Jobless Claims and consumer price index [7]. The traded contracts are securities where payoffs are based on macroeconomic data releases. The instruments are traded as a series (between 10-20) of binary options. For example a single data release of the retail sales in April 2005 was traded as 18 stocks. In order to maximize liquidity the market operators use a series of occasional dutch auctions just before the data releases instead of the more common continuous trading on most financial markets. Thus the market provide hedging opportunities against event risks and a short horizon market forecast of certain economic variables. By analyzing the forecast efficiency Gurkaynak and Wolfers [10] find that market generated forecasts are very similar but more accurate than survey based forecasts[2]. In an attempt to forecast inflation changes in Germany, Berlemann and Nelson [6] set up a series of markets. The markets feature continuous trading of binary contracts. In a similar field experiment Berlemann et al. [5] use a similar system in order to aggregate information about inflation expectations in Bulgaria. All in all, the reported forecasts results in both experiments are mixed but promising.

An Economic Indicator Exchange

In October 2009 a play money prediction market was launched specifically designed to forecast economic indicators such as GDP, inflation, investments, export and unemployment figures in Germany. The goal is to forecast the indicators over longer time periods in advance and continuously aggregate economic information. The market called Economic Indicator Exchange (EIX)[3] was launched in cooperation with the leading German economic newspaper 'Handelsblatt'. The cooperation aims at reaching a wide and well informed audience interested in financial markets and economic development. We thus expect no problems understanding the indicators and the concept of trading. The market is publicly available over the Internet and readers where invited to join. The registration is free and requires besides a valid email address just minimal personal information.

Market and Contract Design

The market design features a continuous double auction without designated market maker. Participants are allowed to submit marketable limit orders with 0.01 increments through the web-based interface. After registration participants are endowed with 1,000 stocks of each contract and 100,000 play money units. We propose to represent continuous outcomes with one stock and define a linear

[2] One must note that the Bloomberg survey forecasts are published on Fridays before the data release, whereas the auction was run -and the forecast was generated- on the data release day.

[3] `www.eix-market.de`

payout function. Contracts for each economic indicator are paid out according to equation 1.

$$p = 100 + \alpha \times (\frac{I_{t0} - I_{t-1}}{I_{t-1}})\ with\ \alpha = 10 \quad (1)$$

A contract is worth: 100 +/- α times the percentage change for an indicator in play money (e.g. a change of 2.1 % results in a price of 121). We set α to 10. Therefore the representable outcome events range from -10% to infinity. To represent the whole outcome range from -100%, α could be set to one. Previous work indicates that market participants find it difficult to estimate minor changes in the underlying event [19]. Hence we propose to scale the minor changes to a certain level. Looking at historical data there were no events where German GDP dropped 10% per quarter. The rationale for setting α to 10 was the deliberation that participants find it more intuitive to enter integers in order to to express reasonable accuracy. Additionally German statistical data releases rarely come with more than one decimal.

Table 1 summarizes the economic variables tradable on the market. Due to the payout function and the selection of the corresponding units; all stock prices are expected to roughly range between 50 and 150. Therefore participants could similarly gain by investing in specific indicators. The indicators are a mix of leading -forecasting the economy- (e.g. Investments) and lagging -describing the state of the economy-(e.g. Unemployment numbers) economic indicators. To facilitate longer forecast horizons every indicator is represented by three independent stocks each representing the next three data releases (t_1, t_2, t_3). As a consequence the initial forecast periods vary between 1 month for monthly released indicators up to 3 quarters for quarterly released variables. One day before the release date the trading in the concerned stock is stopped. Finally the stocks are liquidated according to the payout function defined in equation 1. As soon as the trading in one stock stops a new stock of the same indicator (e.g. t_4) is introduced into the market. This means that participants received 1000 new stocks of the respective indicator. All in all participants are able to continuously trade 18 stocks at all times.

Table 1. Economic variables

Indicator	Unit	Data release cycle	Number of Payouts	Payout function
Exports	$rel. - Changes_{t-1}$	monthly	12	$100 + \alpha \times (\frac{I_{t0} - I_{t-1}}{I_{t-1}})$
GDP	$rel. - Changes_{t-1}$	quarterly	4	$100 + \alpha \times (\frac{I_{t0} - I_{t-1}}{I_{t-1}})$
IFO Index	$abs. - Changes_{t-1}$	monthly	3	$100 + \alpha \times (I_{t0} - I_{t-1})$
Inflation	$rel. - Changes_{t-12}$	monthly	11	$100 + \alpha \times (\frac{I_{t0} - I_{t-12}}{I_{t-12}})$
Investments	$rel. - Changes_{t-1}$	quarterly	5	$100 + \alpha \times (\frac{I_{t0} - I_{t-1}}{I_{t-1}})$
Unemployment	Million (abs.)	monthly	12	$100 + \frac{ABS(Number)}{100.000}$

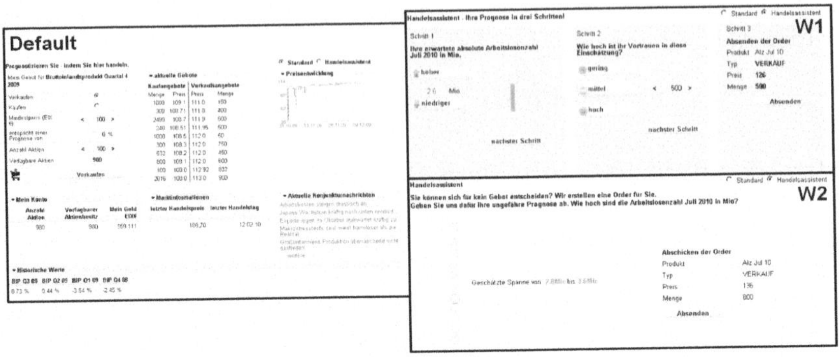

Fig. 1. Three trading interfaces

Incentives

As mentioned the market is a free to join play money market. In order to motivate participants intrinsically we provided two interface features; traders could follow their performance on a leader board and they could form groups with others to spur competition with friends. Previous research in the field of prediction markets has shown that play-money perform as well as real-money markets predicting future events [24,16]. Due to the legal restrictions on gambling the EIX prediction market has to rely on play money. To increase participants' motivation and to provide incentives to truly reveal information we hand out prizes worth 36,000 Euro. As we try to forecast longer periods the incentive scheme has to address this problem. So the incentives are divided in two parts (a) monthly prizes and (b) yearly prizes. The 8 yearly prizes (total value 10,000 Euro) are handed out according to the portfolio ranking at the end of the market. The monthly prizes are shuffled among participants who fulfilled two requirements for the respected month: (i) they increased their portfolio value and (ii) they actively participated by submitting at least five orders. Both incentives are clearly communicated through the interface. For the yearly prizes the leader board indicates the current status of all participants. the monthly winning status is displayed individually just after each login.

Trading Interfaces

The three trading interfaces are displayed in figure 1. In the default trading screen (left side), participants have convenient access to the order book with 10 levels of visible order book depth, the price chart, the account information and market information such as the last trading day. As additional information the Handelsblatt provides access to an up-to-date economic news-stream and finally the indicator's last years performance is displayed. Participants are able to customize their default trading interface individually. By clicking the small arrows the six information panels open and close. In the default setting, only the trading mask and the six headlines are visible. After each submitted order the chosen interface is saved per user. On user return the system opens the previously used interface elements on default. Moreover, a short description of the market comprising the respective payoff function is shown as part of the trading screen.

Additionally to the default trading interface, participants have the choice to switch to a trading wizard guiding their trading decisions. In order to test for the interface influence on trading performance we designed two different wizards displayed on the right hand side, marked with W_1 and W_2. Participants are randomly assigned in one of two groups with access to one of the two different trading wizards. Interface W_1 is designed as a three step trading wizard, with three (green) boxes appearing in order. In the first step participants indicate if they believe the prediction to be higher or lower than the current market forecast. In the second step they are asked about their confidence in their prediction. The third box just displays the generated order. Interface W_2 simply asks the participant to indicate a prediction interval with two handles. On the right hand side an order is automatically generated depending on the current orderbook and the distance between lowest and highest indicated prediction value. The interface is similar to and was inspired by the Yoopick interface [8]. It is noteworthy that both wizards provide far less information than the default interface. In terms of Seuken et al. interface type W_1 can be considered as a weakly hidden market interface, whereas type W_2 hides the market completely [17].

3 Research Model

As more trading decisions are facilitated through (web-based) trading support systems one of the most urging questions is how to design such interfaces. In order to answer this higher research question we have to deeply understand if and how different interfaces influences trading behavior and performance.

To give indications for these research questions we start by analyzing the participants' trading behavior and how the resulting trading performance is influenced by different trading interfaces (Figure 2). We expect users who are familiar with market environments to use the default interface with more information. Users with no market experience might feel confused by too much data and hence reduce the interface to the simple basics. In the first step (H1) we present how the self-chosen interface influences the participants' trading behavior. As all traders have the same start portfolio the size of a trade is a proxy for the trader's confidence perception. Assuming that participants using the wizards are less confident about how to trade, it seems reasonable that the resulting order size is on average lower.

Another individual market behavior is how participants submit their orders. We distinguish between market orders and limit orders. Market orders trade instantaneously against a standing limit order. Therefore the trader submitting a market order pays the effective spread in order to execute directly knowing that the order will be executed. A less confident trader wants to keep the effective spread and posts limit orders. As the wizards do not display the current orderbook, it is reasonable to assume that wizard users are more likely to submit

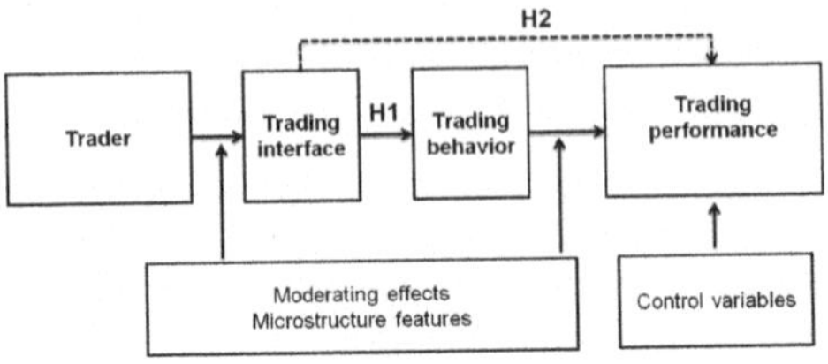

Fig. 2. Research Model

market orders. As a consequence the hypothesis for participants trading behavior (H1) are:

H1a) *Orders which are submitted through a trading wizard are smaller in size on average.*

H1b) *Using the trading wizard increases the chance that participants submit market orders.*

Finally and most importantly, we analyze how the self-chosen interface influences the participants' trading performance (H2). As more information is displayed in the default trading interface an intuitive assumption is to expect a better trading performance through the default interface. However an alternative perspective from decision theory is, that the more information, the worse the performance [15]. Thus the hypothesis for the interface influence on trading performance (H2) are:

H2a) *Using a trading wizard improves the participants' trading performance.*
H2aa) *Using a trading wizard impairs the participants' trading performance.*

In combination the two steps provide a first empirical analysis of the hidden market design paradigm. Moreover they provide insight how a market's interface effect individual trading behavior and subsequently trading performance.

4 Data and Methodology

The following section first presents some descriptive market statistics and then details the tools to systematically analyze the effect of different trading interfaces on trading behavior.

4.1 Descriptive Statistics

The following data includes the timespan from 30th October 2009 till 31st of October 2010. In total 1006 participants registered at the EIX market, of those 680 submitted at least one order. We discard all stocks with less than 50 transactions. Altogether participants submitted 45,808 orders resulting in 22,574 executed transactions. In the respected time frame 47 stocks were paid out. Previous work showed that the market-generated forecasts performed well in comparison to the 'Bloomberg'- survey forecasts, the industry standard. [21]. Out of the

45,808 orders, 821 were submitted through one on the trading wizards. For every order the interface used for order submission is recorded. In the following an interface variable is 1 when the element is used otherwise it is 0, e.g. variable W_1 is 1 if the alternative trading screen W1 is used (see Figure 1). In our field experiment we asked participants to self-assess their market-knowledge and their knowledge of the German economy. These two self-assessment scales combined give us a confidence proxy ($Conf. = 1$).

4.2 Measuring Trading Behavior and Performance

In our continuous market we observe the outcome, i.e. the fundamental value of each stock. Therefore we can ex-post measure the information content of each order. If an order moved the price in the right direction with respect to the final outcome of the stock, it is informed; whereas an order moving the price in the opposite direction with respect to the final outcome price, it is considered uninformed. Thus we present the following score (equation 2) to capture this process. The price of an order o for the stock i is represented as $price_{o,i}$. The fundamental final outcome value of a stock is represented by fv_i. In other words the score rates an order as profitable or not.

$$score_{o_i} = \begin{cases} 1\, if\;\; price_{o,i} \leq fv_i \,\&\, o_{type} = BUY \\ 1\, if\;\; price_{o,i} \geq fv_i \,\&\, o_{type} = SELL \\ 0\, if\;\; price_{o,i} \geq fv_i \,\&\, o_{type} = BUY \\ 0\, if\;\; price_{o,i} \leq fv_i \,\&\, o_{type} = SELL \end{cases} \quad (2)$$

As described in the last section to measure the participant's trading confidence we use two proxies. In order to capture how different interfaces impact the submitted quantity we use the following OLS regression.

$$Quantity_o = \alpha + \beta Wiz. + \gamma Init. + \delta Conf. + \sum_{i=1}^{5} \phi_i M_i \quad (3)$$

$$Quantity_o = \alpha + \beta_1 W_1 + \beta_2 W_2 + \gamma Init. + \delta Conf. + \sum_{i=1}^{5} \phi_i M_i \quad (4)$$

We first compare the differences of the submitted order quantity between the wizards and the regular trading interface. We then relate the quantity to the specific interface used by replacing $Wiz.$ dummy by two dummies W_1 and W_2 one for each trading wizard. As the different indicators exhibit different historic variances, e.g. exports are much more volatile than inflation, we control by adding the market dummy variables $M_1 - M_5$. Similarly, to control for the self-assessed confidence we add a $Conf.$ dummy. The control variables are included in all presented regressions.

For the second proxy we look at how users submit their offers. For an executed trade there are only two possibilities; either an order is a limit order or it is market order. The market order is initializing a trade against a standing limit order. As this is a binary outcome we use a binomial logistic regression. If a trade is initializing, which means it is market making, the dependent variable is 1 otherwise it is 0. We code liquidity taking (initializing) orders with $Init. = 1$.

Equation (5) measures the influence of the interfaces on the probability whether a trade is initializing or passive.

$$\log \frac{\pi_{Init}}{\pi_{Trade}} = \beta Wiz. + \delta Conf. + \sum_{i=1}^{5} \phi_i M_i \tag{5}$$

Finally for the profitability measures we adapt equations (5) the following way; we exchange the dependent variable $\log \frac{\pi_{Init}}{\pi_{Trade}}$ by $\log \frac{\pi_{Score}}{\pi_{Trade}}$.

$$\log \frac{\pi_{Score}}{\pi_{Trade}} = \beta Wiz. + \gamma Init. + \delta Conf. + \sum_{i=1}^{5} \phi_i M_i \tag{6}$$

The dependent variable is the score defined in equation 2 which is 1 for a profit and 0 for a loss. As before we control for different risks in the market categories by adding dummy variables $M_1 - M_5$.

5 Results

In this section we will evaluate how two alternative interfaces support non-sophisticated traders participating in a (complex) prediction market. We show how individual behavior differs depending on the interface used. Controlling for different trading behavior we find that market participants using a trading wizard are more likely to submit profitable orders. Following the presented research model we start by analyzing how trader behavior differs if participants use different interfaces. A common proxy for confidence in a trading environment is the submitted quantity. We assumed that participants using the wizards are less confident about trading, and hence the resulting order size is lower on average. In Table 2 the results for regression 3 (Model A) and 4 (Model B) are depicted. Participants using one of the wizards submit orders with a lower quantity of -858 per order on average. Thus we can accept hypothesis H1a. Separating the effect for certain wizard types (Model B), we see that the result holds -in direction- for both alternative trading interface but only for type W_2 significantly. We assumed that participants using the wizard do not see the orderbook and hence submit more market orders. As the estimates in in Table 3, Model C; show this is the case. Accordingly we accept H1b. Additionally we find that confident traders are more likely to submit limit orders possibly in order to keep the realized spread.

Again looking at the particular influence of each interface we find that the results are due to different behavior supported by the wizard type W_2.

We suggested two alternative hypotheses regarding the interface influence on trading performance. One might intuitively suspect that more information on the default trading interface leads to better trading decisions. Turning to Table 4, Model E; reveals that the chance of submitting a profitable order is higher

Table 2. Influence of trading wizards on submitted quantity. Model A gives the values for the regression (3). The estimates show that if a trading wizard is used, the submitted quantity per order is reduced by 858. The effect is significant for the wizard type W2 (Model B). The market dummies $M_1 - M_5$ are omitted. The superscript 'a' denotes significance at the 0.1%, 'b' at the 1% level and 'c' at the 5% level.

	Wizard	W_1	W_2	Init.	Conf.
Model A	-858[c]	-	-	-53	92
(t-Value)	(-2.2)	(-)	(-)	(0.97)	(1.57)
Model B	-	-742	-911[c]	-53	93
(t-Value)	(-)	(-1.07)	(-1.94)	(-0.97)	(1.58)

Table 3. Influence of trading wizards on order type. The estimates represent the change in the log odds of the outcome if the predictor variable is one. (The chance that an order is a market order is increased if the order is submitted through the trading wizards.) The market dummies $M_1 - M_5$ are omitted. The superscript 'a' denotes significance at the 0.1%.

	Wizard	W_1	W_2	Conf.
Model C	1.38[a]	-	-	-0.37[a]
(χ^2)	(28.33)	(-)	(-)	(124.61)
Model D	-	0.55	1.83[a]	-0.37[a]
(χ^2)	(-)	(1.9)	(27.51)	(125.84)

using one of the wizards. We thus reject hypothesis H2aa and accept H2a. A possible explanation for this result might be that certain information provided by the system may not actually help but impair the trading decision process. Interestingly, looking at how trading behavior relates to successful orders, we see that initializing orders are less likely to be profitable. However without controlling for different market behavior (Table 4; Model E_1) we find that the order submitted through the wizards are still more likely to be profitable. As before it seems that the results are stronger for interface W_2 the strongly hidden market interface.

Table 4. Influence of trading wizards on order profitability. The estimates represent the change in the log odds of of the outcome if the predictor variable is one. (The chance that an order is profitable is increased if the order is submitted through one of the trading wizards.) The market dummies $M_1 - M_5$ are omitted. The superscript 'a' denotes significance at the 0.1%.

	Wizard	W_1	W_2	Init.	Conf.
Model E	0.93[a]	-	-	-0.17[a]	0.18[a]
(χ^2)	(13.3)	(-)	(-)	(28.21)	(29.71)
Model E_1	0.87[a]	-	-	-	0.19[a]
(χ^2)	(11.9)	(-)	(-)	(-)	(35.1)
Model F	-	0.13	1.42[a]	-0.16[a]	0.18[a]
(χ^2)	(-)	(0.11)	(16.05)	(28.6)	(28.86)
Model F_1	-	0.11	1.33[a]	-	0.19[a]
(χ^2)	(-)	(0.77)	(14.59)	(-)	(34.21)

6 Conclusion

In future, various allocation problems call for market based solutions. However market complexities impose high entry barriers for non-sophisticated users. One reason is that in markets preferences are usually communicated through bids and offers which requires participants to adapt to a different mental model. Recently researchers proposed the idea of hidden market design which merges the fields of market design with user interface design in order to make complex markets accessible to a broader audience. As more trading decisions are facilitated through (web-based) 'trading support systems' one of the most urging questions is how to design such interfaces. Moreover it is important to design such interfaces without reducing market efficiency and individual trading performance. In our field experiment participants trade in a complex prediction market which closely reassembles trading in financial markets. As the outcome of events in prediction markets is finally known, we can ex-post measure the participants' trading performance. Evaluating the hidden market paradigm from an individual perspective, we find that alternative trading interfaces change participants' behavior. Using the trading wizards, traders are more likely to submit market orders and submit orders with smaller sizes. Against naive intuition, we find that orders submitted through the strongly hidden market interface are more likely to be profitable compared to orders submitted trough the default trading interface. A reason for that may be found in cognitive theory. Market complexity increases the participants' cognitive load and hence may reduce trading performance and confidence. As a result this work provides insight into the interplay between market design, interface, and trading behavior. Specifically in the domain of financial markets it is the first work to show the influence of the trading interface on trading behavior and performance.

References

1. Ariely, D.: Controlling the information flow: Effects on consumers' decision making and preferences. Journal of Consumer Research: An Interdisciplinary Quarterly 27(2), 233–248 (2000)
2. Arrow, K.J., Forsythe, R., Gorham, M., Hahn, R., Hanson, R., Ledyard, J.O., Levmore, S., Litan, R., Milgrom, P., Nelson, F.D., Neumann, G.R., Ottaviani, M., Schelling, T.C., Shiller, R.J., Smith, V.L., Snowberg, E., Sunstein, C.R., Tetlock, P.C., Tetlock, P.E., Varian, H.R., Wolfers, J., Zitzewitz, E.: Economics: The promise of prediction markets. Science 320(5878), 877–878 (2008)
3. Berg, J.E., Nelson, F.D., Rietz, T.A.: Prediction market accuracy in the long run. International Journal of Forecasting 24(2), 285–300 (2008)
4. Berg, J.E., Rietz, T.A.: Prediction markets as decision support systems. Information Systems Frontiers 5(1), 79–93 (2003); Prediction Markets as Decision Support Systems
5. Berlemann, M., Dimitrova, K., Nenovsky, N.: Assessing market expectations on exchange rates and inflation: A pilot forecasting system for bulgaria. SSRN eLibrary (2005)

6. Berlemann, M., Nelson, F.: Forecasting inflation via experimental stock markets some results from pilot markets (Ifo Working Papers No. 10) (2005)
7. Gadanecz, B., Moessner, R., Upper, C.: Economic derivatives. BIS Quarterly Review (2007)
8. Goel, S., Pennock, D., Reeves, D., Yu, C.: Yoopick: a combinatorial sports prediction market. In: Proceedings of the 23rd National Conference on Artificial Intelligence, vol. 3, pp. 1880–1881. AAAI Press (2008)
9. Goel, S., Reeves, D., Watts, D., Pennock, D.: Prediction without markets. In: Proceedings of the 11th ACM Conference on Electronic Commerce, pp. 357–366. ACM (2010)
10. Gurkaynak, R., Wolfers, J.: Macroeconomic derivatives: An initial analysis of market-based macro forecasts, uncertainty, and risk. NBER Working Papers 11929, National Bureau of Economic Research, Inc. (January 2006)
11. Hahn, R.W., Tetlock, P.C.: Introduction to information markets. In: Hahn, R.W., Tetlock, P.C. (eds.) Information Markets: A New Way of Making Decisions, pp. 1–12. AEI Press, Washington D.C (2006); Incollection George Introduction to Information Markets
12. Hanson, R.: Decision markets. IEEE Intelligent Systems 14(3), 16–19 (1999)
13. Kauffman, R.J., Diamond, L.: The business value effects of cognitive biases in trading workstation window design. SSRN eLibrary (1989)
14. Luckner, S., Schroeder, J., Slamka, C.: On the forecast accuracy of sports prediction markets. In: Gimpel, H., Jennings, N.R., Kersten, G., Ockenfels, A., Weinhardt, C. (eds.) Negotiation, Auctions and Market Engineering, vol. 1, pp. 227–234 (2008)
15. Malhotra, N.K.: Information load and consumer decision making. Journal of Consumer Research: An Interdisciplinary Quarterly 8(4), 419–430 (1982)
16. Servan-Schreiber, E., Wolfers, J., Pennock, D.M., Galebach, B.: Prediction markets: Does money matter? Electronic Markets 14, 243–251 (2004)
17. Seuken, S., Jain, K., Parkes, D.C.: Hidden Market Design. In: Proceedings of the 24th AAAI Conference on Artificial Intelligence (AAAI 2010), pp. 1498–1503 (2010)
18. Seuken, S., Jain, K., Tan, D., Czerwinski, M.: Hidden Markets: UI Design for a P2P Backup Application. In: Proceedings of the Conference on Human Factors in Computing Systems, CHI 2010 (2010)
19. Stathel, S., Luckner, S., Teschner, F., Weinhardt, C., Reeson, A., Whitten, S.: AKX - An Exchange for Predicting Water Dam Levels in Australia. In: Proceedings of the 4th International Symposium on Information Technologies in Environmental Engineering, Thessaloniki, Greece, pp. 78–90 (May 2009)
20. Streufert, S., Driver, M.J., Haun, K.W.: Components of response rate in complex decision-making. Journal of Experimental Social Psychology 3(3), 286–295 (1967)
21. Teschner, F., Stathel, S., Weinhardt, C.: A Prediction Market for Macro-Economic Variables. In: Proceedings of the Forty-Fourth Annual Hawaii International Conference on System Sciences, HICSS (2011)
22. Vessey, I.: The effect of information presentation on decision making: a cost-benefit analysis. Inf. Manage. 27(2), 103–119 (1994)
23. Vessey, I., Galletta, D.F.: Cognitive fit: An empirical study of information acquisition. Information Systems Research 2(1), 63–84 (1991)
24. Wolfers, J., Zitzewitz, E.: Prediction markets. Journal of Economic Perspectives 18(2), 107–126 (2004)

Expectations: Point-Estimates, Probability Distributions, Confidence, and Forecasts

David Rothschild

Yahoo! Research
111 West 40th Street, 17th Floor, New York, NY 10018
David@ResearchDMR.com

Abstract. In this article I test a new graphical, interactive interface that captures both "best estimate" point-estimates and probability distributions from non-experts. As in the previous literature, respondents are overconfident. My innovation is to show that in contrast to the standard method of directly asking respondents to state their confidence, using my method, which induces the respondents to reveal confidence, there is a sizable and statically significant positive relationship between the respondents' confidence and the accuracy of their individual-level expectations. This positive correlation between confidence and accuracy can be utilized to create confidence-weighted aggregated forecasts that are more accurate than the standard "consensus forecasts". The payment of financial incentives does not affect these findings.

Keywords: Polling, information aggregation, belief heterogeneity.

P. Coles et al. (Eds.): AMMA 2011, LNICST 80, p. 18, 2012.

Automated Market Makers That Enable New Settings: Extending Constant-Utility Cost Functions

Abraham Othman and Tuomas Sandholm

Computer Science Department, Carnegie Mellon University
{aothman,sandholm}@cs.cmu.edu

Abstract. Automated market makers are algorithmic agents that provide liquidity in electronic markets. We construct two new automated market makers that each solve an open problem of theoretical and practical interest. First, we formulate a market maker that has bounded loss over separable measure spaces. This opens up an exciting new set of domains for prediction markets, including markets on locations and markets where events correspond to the natural numbers. Second, by shifting profits into liquidity, we create a market maker that has bounded loss in addition to a bid/ask spread that gets arbitrarily small with trading volume. This market maker matches important attributes of real human market makers and suggests a path forward for integrating automated market making agents into markets with real money.

1 Introduction

Automated market makers are algorithmic agents responsible for maintaining open interest in electronic markets, and they have been applied successfully to create new prediction markets that would not exist without the intervention of liquidity-providing agents. For instance, the Gates Hillman Prediction Market successfully forecast a fine-grained probability distribution over potential opening dates for the new computer science buildings at Carnegie Mellon University [Othman and Sandholm, 2010], and Yoopick was a Facebook application that allowed users to predict point spreads with far more detail than in traditional sports betting [Goel et al., 2008]. In this paper, we develop two new automated market makers that can be applied to new domains.

First, we design a loss-bounded automated market maker for separable measure spaces. Automated market making research has centered on finite discrete event spaces, but the market maker we present here is suitable for countably infinite discrete event spaces and uncountably infinite continuous event spaces. This market maker opens up exciting possibilities for valuable new markets. For instance, a prediction market over earthquakes in California or hurricanes in Florida could be run; traders could select ranges of locations and dates and magnitudes (all of these being continuous variables), and get paid if their predictions are correct. More importantly, administrators could use market prices to

P. Coles et al. (Eds.): AMMA 2011, LNICST 80, pp. 19–30, 2012.

get a better picture of disaster risks, providing vital insight that could improve preparation and planning.

We are also interested in expanding the use of automated market makers beyond mediating prediction markets (that typically use artificial currency). We would like to see automated market makers used in real markets with real money. As we saw in the recent "Flash Crash", where open interest in popular equities dropped precipitously [Sterling, 2010], it would be of immense value to have trading regulated by algorithmic agents with predictable behavior. The second market maker we design replicates the desirable features of existing human-mediated markets, but in an algorithmic way. The market maker can run at a profit and is appropriate for use in markets where the level of liquidity is not known *a priori*. Our new market maker reduces its bid/ask spread to zero with sufficient trading volume while still retaining a finite bound on loss. No existing market maker in the literature has both of these properties.

Though the problems we solve sound distinct, our solutions are very similar; both involve extending the same existing framework in different ways. We proceed with an introduction to automated market making, and then discuss each of our new market makers in turn.

2 Setting

To construct an automated market maker, the space of possible futures is exhaustively partitioned into a set of events Ω, such that exactly one of the $\omega \in \Omega$ will be realized. The bets traders make with the market maker are represented as functions from the event space to the real numbers, so that bets are contracts specifying how much the market maker will pay out if each of the various futures is realized. Typically, automated market makers use *cost functions* to price trades.

Definition 1. *Over n events, a* cost function *is a scalar field $C : \Re^n \mapsto \Re$ that maps a vector of amounts the market maker must pay out (the* obligation vector*) to a scalar value (the* cost*).*

The automated market maker keeps track of their obligation vector over the sequence of bets traders make. A trader who wishes to move the obligation vector from $\mathbf{x}$ to $\mathbf{y}$ is charged $C(\mathbf{y}) - C(\mathbf{x})$.

(In our notation, x is a scalar and $\mathbf{x}$ is a vector. x_i refers to the i-th element of a discrete vector, and $\mathbf{x}(t)$ refers to the evaluation of a vector function at point t (when we move to general vector spaces). ∇f is the gradient of the function f and $\mathbf{1}$ refers to a vector in which every component is 1.)

Over a series of transactions, the cost function telescopes from some initial $\mathbf{x^0}$ (which is often just set to $\mathbf{0}$). Consequently, the scalar difference between the current payout vector $\mathbf{x}$ and the initial payout vector $\mathbf{x^0}$, $C(\mathbf{x}) - C(\mathbf{x^0})$, is the amount traders have paid to the automated market maker. When an event ω_i is realized, the market maker's profit or loss is $C(\mathbf{x}) - C(\mathbf{x^0}) - x_i$. If, for all $\mathbf{x}$ and all ω_i this quantity is bounded, we say the market maker has *bounded worst-case loss* (or just *bounded loss*). A review of the mechanics of automated market makers can be found in Pennock and Sami [2007].

Most automated market makers use cost functions that are *convex risk measures*:

Definition 2. *A* convex risk measure *is a cost function C that satisfies three properties:*

Monotonicity If $\mathbf{x} \leq \mathbf{y}$*, then* $C(\mathbf{x}) \leq C(\mathbf{y})$.
Convexity For all $\mathbf{x}$ *and* $\mathbf{y}$*, and* $\lambda \in [0, 1]$

$$C(\lambda\mathbf{x} + (1 - \lambda)\mathbf{y}) \leq \lambda C(\mathbf{x}) + (1 - \lambda)C(\mathbf{y}).$$

Translation Invariance For all $\mathbf{x}$ *and* α*,* $C(\mathbf{x} + \alpha\mathbf{1}) = C(\mathbf{x}) + \alpha$.

Monotonicity ensures that traders cannot profit without taking on some risk, convexity avoids pricing degeneracies, and translation invariance means that buying or selling a guaranteed profit of a dollar costs a dollar.

Examples of convex risk measures in the literature include the LMSR [Hanson, 2003, 2007], sequential convex parimutuel mechanisms [Peters et al., 2007; Agrawal et al., 2009], market makers derived from no-regret online learning algorithms [Chen et al., 2008; Chen and Vaughan, 2010], and constant-utility cost functions [Chen and Pennock, 2007], the notion we extend in this work.

One alternative to convex risk measures is the liquidity-sensitive cost function of Othman et al. [2010], which we discuss later in this paper.

Market makers that do not use cost functions at all include heuristic market makers [Das, 2008; Brahma et al., 2010] (which do not rule out manipulability), subsidized dynamic pari-mutuel mechanisms [Pennock, 2004] (which do not offer guaranteed payoffs), and self-financing static pari-mutuel mechanisms [Lambert et al., 2008] (which do not offer guaranteed payoffs or sequential interactions with a market maker). We will not discuss these alternatives to cost-function-based market makers further in this paper.

2.1 Constant-Utility Cost Functions

In this section we describe *constant-utility cost functions*, an existing framework for building cost functions that was introduced by Chen and Pennock [2007]. Our new market makers are created by extending this framework.

Definition 3. *A* utility function *is an increasing concave function* $u : \Re \mapsto \Re$.

A constant-utility cost function works by charging traders the amount that keeps the market maker's utility at a constant level. Put another way, the market maker prices each bet so that he is indifferent between a trader declining and accepting it.

Definition 4. *Let* $x^0 \in \operatorname{dom} u$ *and let* μ *be a probability distribution over the* ω_i*, defining* $p_i \equiv \mathbb{P}_\mu(\omega_i)$*. A* constant-utility cost function $C : \Re^n \mapsto \Re$ *is defined implicitly as the solution to*

$$\sum_i p_i u(C(\mathbf{x}) - x_i) = u(x^0)$$

Since the cost function is given implicitly by a root-finding problem, rather than explicitly as a function of the vector $\mathbf{x}$, it is not immediately clear that costs exist, are unique, and are easily computable. However, because the utility function is monotonically increasing, the cost function exists for any input vector and is unique. Furthermore, because the utility function is increasing, we can compute b bits of the value of the cost function in b steps of a binary search over possible values.

The partial derivatives of the cost function are known as *prices*. Where u is differentiable the prices of a constant-utility cost function have a closed form [Jackwerth, 2000; Chen and Pennock, 2007]:

Where u is differentiable the prices of a constant-utility cost function form a probability distribution:

$$\nabla_i C(\mathbf{x}) = \frac{p_i u'(C(\mathbf{x}) - x_i)}{\sum_j p_j u'(C(\mathbf{x}) - x_j)}.$$

The prices form a probability distribution because they sum to one and are non-negative (the utility function is non-decreasing and so its derivative is never negative).

We are particularly interested in a special class of utility functions that we dub *log-like*.

Definition 5. *A* log-like *utility function u is a differentiable utility function defined over* $(0, \infty)$ *that is unbounded from below from the right:* $\lim_{x \to 0^+} u(x) = -\infty$.

One example of a log-like utility function is log, and another is $-1/x$. The key property of log-like utility functions that we exploit is the extreme negative weight on input values close to zero.

3 Market Making over Measure Spaces

In this section we extend constant-utility cost functions so that they apply over arbitrary measure spaces, and demonstrate how a particular class of utility functions over separable measure spaces has bounded loss. What makes this challenging is the tension between loss boundedness and sometimes offering traders bets that would be irrational to accept regardless of what event materializes. It is easy to create market makers with bounded loss if they charge agents as much as they could possibly win. On the other hand, it is also easy to create market makers that only offer bets a rational agent could accept, but have unbounded loss. For instance, the most popular automated market maker in practice is the LMSR, which for n events has worst-case loss $\Theta(\log n)$. Over infinite event spaces, however, it has unbounded loss [Gao et al., 2009].

Because of this tension, it has been an open problem whether there exist market makers with bounded loss over arbitrary measure spaces that never offer agents bets which are immediately bad. In this section, we construct such a market maker. Our only restriction is that the spaces must be separable, a technical condition that is met naturally by all spaces of interest—such as the real line or the natural numbers, as well as geometric shapes like cubes and spheres.

3.1 Measure Spaces and Separability

We generalize the usual notation from settings with a finite number of events to settings with potentially an infinite number of events. In this context, the events are $t \in \Omega$, and bets still map from events to how much the trader gets paid based on what event materializes. However, bets and payout vectors (which were represented as points in $\Re^n$ in the finite case) are now functions over the event space; formally, $\mathbf{x} : \Omega \mapsto \Re$.

The cost function from a finite-event setting now generalizes to be a functional that maps these functions $\mathbf{x}$ to scalar values, i.e., $C : (\Omega \mapsto \Re) \mapsto \Re$. The notational similarity with the traditional setting that has only a finite number of events can be seen later because sums are simply replaced by integrals and probability mass functions are replaced by probability density functions.

We assume that the event space is *separable*, meaning that it possesses a natural notion of distance. With a separable event space we can assert the existence of a probability distribution over the events that has positive measure for every event, even though the space of events could be infinite. Formally, the following result is well known in measure theory.

Proposition 1. *In any separable measure space, there exists a probability measure μ such that $\mu(t) > 0$ for all t.*

We call such a μ an *always-positive measure.* Over the real line, an example of an always-positive measure is a Gaussian distribution. Measure spaces that are not separable are highly degenerate and are unlikely to be of practical importance.

3.2 Market Making without Offering Bad Bets

The aim of this section is to devise a market maker that has bounded loss over separable measure spaces that does not offer agents bets which are immediately bad. This latter quality is important because it is simple to construct a market maker with bounded loss over arbitrary measure spaces. The sup cost function (which can be thought of as an arbitrary-dimensional max)

$$C(\mathbf{x}) = \sup_t \mathbf{x}(t)$$

accomplishes this goal. This is easy to see because a trader is charged as much as she could possibly gain from the realization of any event. However, sup is one example of a cost function that *offers bets which are immediately bad.*

Definition 6. *A differentiable convex risk measure C* offers bets which are immediately bad *if there exists an obligation vector $\mathbf{x}$ where a proper subset of the event space has prices that sum to at least 1.*

As the name implies, trade with a cost function of this sort can sometimes be a dominated action: a trader pays at least as much as she could possibly make from any contract, but there also exist events for which the trader loses her bet, too.

Agrawal et al. [2009] present several cost functions with a worst-case loss of $\Theta(1 - 1/n)$, which implies bounded loss as n gets large. However, they achieve their bounded worst-case loss by offering good bets only up to a certain amount of uncovered exposure. After reaching this limit (which is controlled by a parameter set *a priori* by the market administrator) the price of some events is set to zero, implying that the other events form a proper subset with unit marginal price. This makes the market maker offer bets which are immediately bad. (Furthermore, the market makers in Agrawal et al. [2009] rely on solving a relatively complex convex optimization, so it is not immediately clear how to generalize their technique over infinite-dimensional spaces, or whether such a generalization is even possible.)

3.3 Our Loss-Bounded Market Makers

We now show that using a constant-utility cost function with a log-like utility function and an always-positive measure produces a market maker with bounded loss over separable measure spaces that never offers bets to agents which are immediately bad. These are the first market makers to have both of these properties.

These market makers could be used over continuous spaces, for instance, to create time and location markets for natural disasters like earthquakes, tornadoes, and hurricanes. Furthermore, events that take place over unbounded discrete sets (such as population sizes, interest rates, or stock prices) can also be handled effectively with these market makers because countable sets are also separable measure spaces.

Our market maker requires solving an integral equation over the measure space, and so the feasibility of our scheme relies on how easy it is to evaluate the integral. We suggest that while it is possible to describe degenerate cases where it is not possible to compute this integral effectively, for natural domains like geometric shapes or number lines the integral can be computed using numerical techniques in a straightforward manner.

Proposition 2. *Let u be a log-like utility function, μ be an always-positive measure, and $x^0 > 0$. Then the constant-utility cost function given by the solution to*

$$\int_t u(C(\mathbf{x}) - \mathbf{x}(t))\mu(t)\,dt = u(x^0)$$

loses at most x^0 and never offers bets which are immediately bad.

Proof. To prove the bounded loss property, it suffices to show that there exists no $\mathbf{x}$ and t such that $\mathbf{x}(t) - C(\mathbf{x}) > x^0$, because if such an $\mathbf{x}$ and t existed it would imply that by moving the market maker to the payoff vector $\mathbf{x}$ and realizing the event t the market maker would lose more than x^0. Suppose there did exist such an $\mathbf{x}$ and t. By definition, evaluating the integrand at t would involve evaluating

$$u(C(\mathbf{x}) - \mathbf{x}(t)) = -\infty$$

because the utility function is log-like. Since the probability measure is always positive, $\mu(t) > 0$ and so the resulting integral is undefined rather than $u(x^0)$, a contradiction.

To prove the resulting cost function never offers bets which are immediately bad, consider that over a separable measure space a differentiable convex risk measure offer immediately bad bets if there exists a proper subset of the space A and an obligation vector $\mathbf{x}$ such that

$$\int_A \nabla C(\mathbf{x})\mu(t)\, dt = 1$$

while

$$\int_A \mu(t)\, dt \neq 1$$

For there to not exist such an $A, \mathbf{x}$ pair, it is suffices that for all $\mathbf{x}$ and t, $\nabla C(\mathbf{x})(t)\mu(t) > 0$ or, because μ is an always positive measure $\nabla C(\mathbf{x})(t) > 0$ for all $\mathbf{x}$ and t.

Since C is a constant-utility cost function, recall the closed-form expression for its gradient (which over general measure spaces is a function)

$$\nabla C(\mathbf{x})(t) = \frac{\mu(t)u'(C(\mathbf{x}) - \mathbf{x}(t))}{\int u'(C(\mathbf{x}) - \mathbf{x}(t))\mu(t)\, dt}$$

Because the utility function is strictly increasing and differentiable, its derivative exists and is always positive. Coupling this with the fact that μ is an always positive measure, we have that the numerator

$$\mu(t)u'(C(\mathbf{x}) - \mathbf{x}(t)) > 0$$

and so $\nabla C(\mathbf{x})(t) > 0$ for all $\mathbf{x}$ and t. Consequently, there never exists a proper subset A of the space that has unit marginal price. Thus, the cost function never offers bets that are immediately bad. ■

4 Funneling Profits into Liquidity

In this section we construct a cost-function-based market maker with a fixed worst-case loss that also shrinks its bid/ask spread to zero (provided there is sufficient transaction volume). In these characteristics, and also in its ability to realize a profit, the market maker we describe in this section closely matches human market makers we see operating in the real world. This suggests that our automated market maker could prove to be a practical choice for markets with real money.

We achieve these properties through realizing the notion of "driving profits toward liquidity", an idea that has never been successfully and fully implemented in an automated market maker. Probably the closest prior effort is the market maker of Othman et al. [2010], which features a liquidity-sensitive market maker with a shrinking bid/ask spread. However, that market maker can only shrink its bid/ask spread to a positive constant determined *a priori*, not arbitrarily small.

4.1 Description of the Problem

Liquidity is the amount of active interest in a market. It is intimately related to the *bid/ask spread*, or the difference between the price to buy and the price to sell a contract. Real world markets are characterized by an inverse relationship between transaction volumes and bid/ask spreads. *Deep* (aka *liquid*) markets have small bid/ask spreads; for instance, large cap equities generally have one-cent bid ask spreads for thousands of dollars of interest. These markets can sustain large trades without prices moving significantly. In contrast, a *shallow* (aka *illiquid*) market is characterized by large bid/ask spreads and prices that move significantly in response to trades. At an extreme, a very illiquid market may not have any open interest at all.

From an automated market making perspective, there exists a sharp tension between the sensitivity of the market maker to liquidity and the amount that the market maker can lose. Consider the case of a market maker that prices bets according to some probability distribution over the future state of the world, and does not change that distribution in response to the bets placed by traders. Such a market maker always has a bid/ask spread of zero, but also has unbounded worst-case loss. In contrast, the bounded-loss convex risk measures in the literature always have non-vanishing, positive bid/ask spreads for fixed-size bets.

We can also come up with more sophisticated variations on these schemes in an effort to shrink bid/ask spreads, but these also do not satisfy our desiderata. Consider a market maker operating based on a fixed probability distribution who also takes a diminishing profit cut, a notion we will mathematically formalize later in this section. With such a scheme, the bid/ask spread falls to zero with increasing liquidity, but the market maker still has unbounded worst-case loss. On the other hand, a standard bounded-loss convex risk measure equipped with a diminishing profit cut has bounded worst-case loss and a shrinking bid/ask spread, but this spread never goes away even as liquidity gets large.

What has not been resolved is the design of a market maker that (1) has bounded loss and (2) has a bid/ask spread that becomes arbitrarily small with increasing liquidity. In this section, we build such a market maker.

4.2 Diminishing Profit Cuts

In this section we introduce a framework for a market maker adding an additional charge to trades on top of a cost function. While it is easy to, for example, charge a trader 10 cents extra per transaction, making a profit cut that vanishes with liquidity requires more sophistication.

Recall that a constant-utility cost function solves

$$\sum_i p_i u(C(\mathbf{x}) - x_i) = u(x^0)$$

Now consider how to add profits into this calculation. The *profit function* (or *cut*) $\Pi : \Re_+ \mapsto \Re_+$, is a continuous increasing function that keeps track of the amount of profit we have collected. Π relies on the current state s; at the market's initiation we set $s = 0$, and since Π tracks profit, $\Pi(0) \equiv 0$. When the obligation vector of the market moves from $\mathbf{x}$ to $\mathbf{y}$ we set $s = s + ||\mathbf{y} - \mathbf{x}||$.

The following algorithm details how our market maker prices trades as the sum of the cost function and the profit function.

1. We are currently at obligation vector $\mathbf{x}$ and profit state s. A trader wishes to take on a transaction that would leave the market maker with the obligation vector $\mathbf{y}$.
2. The value of the cost function, $C(\mathbf{y})$, is solved through

$$\sum_i p_i u(C(\mathbf{y}) - y_i + \Pi(s + ||\mathbf{y} - \mathbf{x}||)) = u(x^0)$$

   (In this equation, $C(\mathbf{y})$ is the variable and all of the other values are constants.)
3. The total transaction cost quoted to the trader (the *quoted price*) is

$$C(\mathbf{y}) - C(\mathbf{x}) + \Pi(s + ||\mathbf{y} - \mathbf{x}||) - \Pi(s)$$

4. If the trade is accepted, the current profit state is updated, $s \doteq s + ||\mathbf{y} - \mathbf{x}||$.

In order to design our new market maker, we require our profit function to meet some additional requirements. First, it must not have a bound on the total amount it collects $\lim_{s\to\infty} \Pi(s) = \infty$. Second, the marginal profit cut must vanish as liquidity gets large; for differentiable Π this condition is $\lim_{s\to\infty} \Pi'(s) = 0$.

There are many functions that satisfy these conditions. Any such function would constitute a market maker in our desired class, and the propositions in the rest of this section would go through. However, in what follows, we will focus on one particularly simple such function: $\Pi(s) = \alpha \log(s + 1)$ for some $\alpha > 0$. The corresponding marginal profit cut is $\Pi'(s) = \frac{\alpha}{s+1}$.

Equipped with this profit side, we can now introduce our market maker.

4.3 Our New Market Maker

Our market maker uses a constant-utility cost function with the profit cut framework we described above. The utility function is $u(x) = \log(x)$ and the profit function is $\Pi(s) = \alpha \log(s + 1)$.

The market administrator has two parameters to set: the $x^0 > 0$ chosen to initially seed the constant utility level, and the $\alpha > 0$ chosen to moderate the profit function. Higher x^0 leads to higher worst-case loss but more initial liquidity-sensitivity. Higher α makes the profit cut larger.

Our market maker has loss bounded by its initial cost seed:

Proposition 3. *Let* $\min_i p_i > 0$*, so the probability mass function is always positive. Then our market maker loses no more than the* x^0 *chosen to initially seed the constant utility level* $\log x^0$.

Proof. Because we are using a log-like utility function and the probability mass function is never negative, the costs, profits, and obligation vector always obey the relationship $C(\mathbf{x}) + \Pi(s) > \max_i x_i$ because if this were not the case the sum used to calculate expected utility would be undefined. But our loss if the market expires at state $(\mathbf{x}, s)$ is at most

$$\max_i x_i + C(\mathbf{0}) - C(\mathbf{x}) - \Pi(s)$$

Consequently, loss can be no greater than $C(\mathbf{0})$, which for constant-utility cost functions is just x^0 by definition. ∎

Our market maker's bid/ask spread becomes arbitrarily small with increasing activity:

Proposition 4. *The bid/ask spread of our market maker becomes arbitrarily small as* $s \to \infty$.

Proof. We prove that as $s \to \infty$, the quoted price of any fixed trade can be arbitrarily well-approximated by a linear extrapolation from the marginal prices of the cost function. Since the cost function is a differentiable convex risk measure, these marginal prices sum to unity, and so the bid/ask spread becomes arbitrarily small as both the buy and sell sides approach the same quoted prices.

Let the market maker be in state $(\mathbf{x}, s)$, let $\mathbf{y}$ be some trade, and let C_a denote a logarithmic constant-utility cost function seeded with initial cost a. Since profits are funneled into liquidity, we have that the quoted price to move to state $(\mathbf{x} + \mathbf{y}, s + ||\mathbf{y} - \mathbf{x}||)$ is no larger than

$$C_{x^0+\Pi(s)}(\mathbf{x} + \mathbf{y}) - C_{x^0+\Pi(s)}(\mathbf{x}) + \Pi(s + ||\mathbf{y} - \mathbf{x}||) - \Pi(s)$$

Now consider each pair of these terms in turn as s grows large. For the cost function terms, a logarithmic constant-utility market maker with large initial cost becomes increasingly insensitive to the second-order effects (i.e., changing prices) when taking on a fixed size bet. This is because utility grows arbitrarily large in initial cost, while marginal utility shrinks arbitrarily small in initial cost. Consequently, since Π gets arbitrarily large in s we have that for large s

$$C_{x^0+\Pi(s)}(\mathbf{x} + \mathbf{y}) - C_{x^0+\Pi(s)}(\mathbf{x}) \approx \nabla C_{x^0+\Pi(s)}(\mathbf{x}) \cdot \mathbf{y}$$

or, that the difference in costs can be arbitrarily well-approximated by a linear extrapolation from marginal prices.

As for the Π terms, by construction the derivative of the profit cut Π' gets arbitrarily small in s, so for large s the difference of the two terms is arbitrarily small. Furthermore, the upper bound on the quoted price of

$$C_{x^0+\Pi(s)}(\mathbf{x} + \mathbf{y}) - C_{x^0+\Pi(s)}(\mathbf{x}) + \Pi(\mathbf{x} + \mathbf{y}) - \Pi(\mathbf{x})$$

gets arbitrarily tight as s gets large, because as profits get arbitrarily small, the impact of adding them back into the cost function of our market maker gets arbitrarily small, too.

Putting all of these arguments together, the quoted price can be approximated arbitrarily well as s gets large by $\nabla C_{x^0+\Pi(s)}(\mathbf{x})\cdot\mathbf{y}$, which is a linear extrapolation of the marginal prices of the cost function. ■

Acknowledgments. This material is based upon work supported by the National Science Foundation under grants IIS-0905390, IIS-0964579, and CCF-1101668.

References

Agrawal, S., Delage, E., Peters, M., Wang, Z., Ye, Y.: A unified framework for dynamic pari-mutuel information market design. In: ACM Conference on Electronic Commerce (EC), pp. 255–264 (2009)

Brahma, A., Das, S., Magdon-Ismail, M.: Comparing Prediction Market Structures, With an Application to Market Making. Technical report, Dept. of Computer Science, Rensselaer Polytechnic Institute (2010)

Chen, Y., Pennock, D.M.: A utility framework for bounded-loss market makers. In: Proceedings of the 23rd Annual Conference on Uncertainty in Artificial Intelligence (UAI), pp. 49–56 (2007)

Chen, Y., Vaughan, J.W.: A new understanding of prediction markets via no-regret learning. In: ACM Conference on Electronic Commerce (EC), pp. 189–198 (2010)

Chen, Y., Fortnow, L., Lambert, N., Pennock, D.M., Wortman, J.: Complexity of combinatorial market makers. In: ACM Conference on Electronic Commerce (EC), pp. 190–199 (2008)

Das, S.: The effects of market-making on price dynamics. In: International Conference on Autonomous Agents and Multi-Agent Systems (AAMAS), pp. 887–894 (2008)

Gao, X., Chen, Y., Pennock, D.M.: Betting on the real line. In: International Workshop On Internet And Network Economics (WINE), pp. 553–560 (2009)

Goel, S., Pennock, D., Reeves, D., Yu, C.: Yoopick: a combinatorial sports prediction market. In: Proceedings of the National Conference on Artificial Intelligence (AAAI), pp. 1880–1881 (2008)

Hanson, R.: Combinatorial information market design. Information Systems Frontiers 5(1), 107–119 (2003)

Hanson, R.: Logarithmic market scoring rules for modular combinatorial information aggregation. Journal of Prediction Markets 1(1), 1–15 (2007)

Jackwerth, J.: Recovering risk aversion from option prices and realized returns. Review of Financial Studies 13(2), 433 (2000)

Lambert, N., Langford, J., Wortman, J., Chen, Y., Reeves, D., Shoham, Y., Pennock, D.: Self-Financed Wagering Mechanisms for Forecasting. In: ACM Conference on Electronic Commerce (EC), pp. 170–179 (2008)

Othman, A., Sandholm, T.: Automated market-making in the large: the Gates Hillman prediction market. In: ACM Conference on Electronic Commerce (EC), pp. 367–376 (2010)

Othman, A., Pennock, D.M., Reeves, D.M., Sandholm, T.: A practical liquidity-sensitive automated market maker. In: ACM Conference on Electronic Commerce (EC), pp. 377–386 (2010)

Pennock, D., Sami, R.: Computational Aspects of Prediction Markets. In: Algorithmic Game Theory, ch. 26, pp. 651–674. Cambridge University Press (2007)

Pennock, D.M.: A dynamic pari-mutuel market for hedging, wagering, and information aggregation. In: ACM Conference on Electronic Commerce (EC), pp. 170–179 (2004)

Peters, M., So, A.M.-C., Ye, Y.: Pari-Mutuel Markets: Mechanisms and Performance. In: Deng, X., Graham, F.C. (eds.) WINE 2007. LNCS, vol. 4858, pp. 82–95. Springer, Heidelberg (2007)

Sterling, B.: Flash Crash (2010), http://www.wired.com/beyond_the_beyond/2010/05/flash-crash/

Instructor Rating Markets

Mithun Chakraborty, Sanmay Das, Allen Lavoie,
Malik Magdon-Ismail, and Yonatan Naamad

Rensselaer Polytechnic Institute, Troy NY 12180, USA

Abstract. We describe the design of *Instructor Rating Markets* in which students trade on the ratings that will be received by instructors, with new ratings revealed every two weeks. The markets provide useful dynamic feedback to instructors on the progress of their class, while at the same time enabling the controlled study of prediction markets where traders can affect the outcomes they are trading on. More than 200 students across the Rensselaer campus participated in markets for ten classes in the Fall 2010 semester. We show that market prices convey useful information on *future* instructor ratings and contain significantly more information than do past ratings. The bulk of useful information contained in the price of a particular class is provided by students who are in that class, showing that the markets are serving to disseminate insider information. At the same time, we find little evidence of attempted manipulation of the liquidating dividends by raters. The markets are also a laboratory for comparing different microstructures and the resulting price dynamics, and we show how they can be used to compare market making algorithms.

We present a novel application of prediction markets to instructor evaluations. Such markets have the potential to provide dynamic feedback on the progress of a class. We describe a pilot deployment of these markets at Rensselaer Polytechnic Institute in the Fall of 2010, with more than 200 students participating across 10 classes. These markets provide insights into the behavior of students in their roles as both traders and, potentially, as market manipulators (traders who are in a class directly affect the rating of that class), while also allowing us to study how market microstructure affects price formation and the information content of prices.

Each instructor-course pair is an openly traded security in the IRM. Every two weeks, each security pays a liquidating dividend derived from how students in the class rate the instructor for that two week period. Each security can be traded by anyone at the institute, but only students who are in the instructor's class may rate the instructor. A rating period opens after the first week of trading, and students who have "in class" credentials receive an email asking them to rate the instructor of their class – the rating period stays open until the end of the second week, at which point both the rating and trading windows close. If everything works well, fluctuations in the price of the "instructor security" give *real-time feedback* on how well the instructor is doing (we are not endorsing teaching to maximize "stock value", but instructor ratings exist, and it is useful to know

P. Coles et al. (Eds.): AMMA 2011, LNICST 80, pp. 31–32, 2012.

more about what goes into student ratings, and how they would change on a day-to-day basis if students were "polled" repeatedly). Thus, we use students, as well as their roommates and friends, as information gatherers, giving them an outlet (a fun trading game) to reveal their information. While the instructor is only rated occasionally, price movements provide continuous feedback.

There are two major differences between the IRMs and more traditional prediction markets. First, in many prediction markets, information revelation continues right up to the moment of liquidation (for example, opinion polls are released continuously during election cycles), whereas in our markets the only major information revelation is the liquidation event itself. The information revelation leading up to liquidation in IRMs is considerably more noisy (Did the instructor give a good lecture? Was there a hard homework due that week?). Second, typical large prediction markets, such as election markets, attempt to predict a much more stable statistical aggregate quantity: voting turnouts range from the tens of thousands to the tens or hundreds of millions. In contrast, the classes the IRMs ran on in our deployment had between 3 and 25 regular raters. This raises questions about the effects of insider information and potential market manipulation. The success of the markets in predicting instructor ratings is not a given.

However, we find that prices are, in fact, predictive of future instructor ratings, and significantly more predictive than are previous ratings, showing that they incorporate new information. The higher predictivity is due to the trades of insiders: our data shows that when previous and future liquidations differ, students who are enrolled in a class trade in the direction of future liquidations while others trade in the direction of the last liquidation. We also find little evidence of efforts by students to manipulate the ratings for their own benefit as traders: first, the ratings had very high correlation with the official end-of-semester student evaluations of the classes, and, second, we found few cases where students, either individually or in groups, gave surprising ratings and profited from doing so. The fact that IRM ratings are well aligned with the official end-of-semester evaluations shows that the system as a whole is relevant and useful to instructors. Combining that fact with the power of prices to predict IRM ratings is encouraging for the potential of such markets.

In addition to our primary results, we also document learning behavior along several dimensions. In particular, prices for more predictable securities become more efficient, and an early "in class" optimistic bias in traded prices disappears in later periods. The markets also have other beneficial side effects: for example, active traders are more likely to give ratings, thus providing instructors with useful feedback every two weeks. This is already an achievement over the considerably less dynamic single end-of-semester ratings typically available. Finally, we can use the IRM to study the effects of different market microstructures. In particular, we provide further validation of a Bayesian market-making algorithm, BMM, that can provide more price stability than the standard Logarithmic Market Scoring Rule (LMSR) market maker while also making more profit.

An Experimental Study of Sponsored-Search Auctions*

Yeon-Koo Che[1], Syngjoo Choi[2], and Jinwoo Kim[3]

[1] Columbia University and YERI,
Department of Economics, Columbia University, 420 West 118th St, 10161AB, New York, NY 10027, USA
yc2271@columbia.edu
http://www.columbia.edu/~yc2271

[2] UCL,
Department of Economics, University College London, Gower Street, London WC1E 6BT, UK
syngjoo.choi@ucl.ac.uk
http://www.homepages.ucl.ac.uk/~uctpsc0

[3] Yonsei University,
Department of Economics, Yonsei University, 262 Seongsan-No, Seodaemum-Gu, Seoul 120-749, Korea
jikim72@gmail.com
http://web.yonsei.ac.kr/jikimdir

Abstract. We study the Generalized Second Price auctions—a standard method for allocating online search advertising—experimentally, considering both the static environment assumed by the prevailing theory and a dynamic game capturing the salient aspects of real-world search advertising auctions. We find that subjects tend to overbid in both treatments relative to the Vickrey-Clarke-Groves outcome suggested as most plausible by the theory, but that their behavior in the dynamic game resembles the behavior in the static game. Our analysis thus lends support to the use of a static game as modeling proxy, but calls into question the prevailing equilibrium predictions.

JEL Classification: C92, D44, M3.

Keywords: online advertising, sponsored search auction, generalized second price auction, experiment.

---

* We are grateful to Jacob and Michelle Goeree, John Kagel, Dan Levin, Michael Ostrovsky, Michael Schwarz, and the participants at the Ohio State University, University of Zurich and the WCU Market Design Conference held at Yonsei University in August 2010, for valuable comments, and to Brian Wallace for writing the experimental program, and to Tom Rutter for helping us run the experiment. Che and Kim acknowledge the support by WCU program through the National Research Foundation of Korea funded by the Ministry of Education, Science and Technology (R32-2008-000-10056-0). Choi acknowledges the Economic and Social Research Council (UK) Grant No. RES-061-25-0348 and via the ELSE for financial support.

P. Coles et al. (Eds.): AMMA 2011, LNICST 80, p. 33, 2012.

Sharing in BitTorrent Can Be Rational
(Extended Abstract)*

Mike Ruberry and Sven Seuken

Harvard University,
Cambridge, MA 02138, USA
{mruberry,seuken}@eecs.harvard.edu

1 Introduction: BitTorrent Is a Goal-Oriented Game

Consider a game played by two mountaineers climbing a mountain. Both are only interested in attaining the summit and can only reach it with help from the other. Over an infinite number of discrete periods they play a symmetric simultaneous game where they may either help the other a fixed distance up the mountain at some cost, or do nothing. Can these mountaineers climb their mountain?

BitTorrent, the popular peer to peer file distribution protocol, is strategically similar to this mountain climbing game. Peers' single goal is to acquire the complete file—to reach the summit—and incomplete files have zero value. Also, like mountaineers, peers can only progress with others' help. Of course, a BitTorrent swarm is more complex than the mountaineering game, containing many peers who can only provide certain pieces of the file to a subset of others. Importantly, however, this metaphor captures BitTorrent's *goal-oriented* nature, an often overlooked but salient feature of BitTorrent.

In our paper we explicitly model BitTorrent as a *goal-oriented game* that has a unique one time payoff for acquiring the complete file. We use this model to obtain novel results about the rationality of sharing in BitTorrent, distinct from previous work on peer to peer networks (e.g. [1] and [3]) and repeated matching games [4]. Only by accurately modeling peers' incentives and understanding their rational strategies can we design maximally efficient file distribution protocols.

2 Summary of Results

We define a *BitTorrent game* to study the rationality of strategies (protocols). We use *rational* in the formal game theory sense of the word, a weak constraint informally requiring that peers act to maximize their utility. While people are not always rational, they are likely to choose "rational" software offering shorter download times and consuming less upload bandwidth.

Result 1: *In the basic BitTorrent game, sharing is not rational.*

* Working paper available at:
www.seas.harvard.edu/~mruberry/RationalBitTorrent2011.pdf

P. Coles et al. (Eds.): AMMA 2011, LNICST 80, pp. 34–35, 2012.

The intuition for this negative result is that in the basic BitTorrent game, a rational agent will not provide a peer the last piece of the file it needs. This is because peers that obtain all the pieces of a file immediately leave the swarm—they reach the mountain top and are uninterested in helping you up. This "last piece problem" causes a breakdown of collaboration throughout the system, even when the pieces held by other peers are unobserved. In real world BitTorrent swarms, however, some peers are altruists who enjoy providing pieces of the file to assist its distribution. When altruists are present, collaboration can be established:

Result 2: *In the BitTorrent game with altruists, sharing can be rational.*

The presence of altruists allows selfish peers to masquerade as altruists, creating *type uncertainty.* If a peer fails to share with an altruist while it is acquiring the file, the altruist will not provide pieces for free later. With selfish and altruistic peers indistinguishable, sharing becomes rational when peers are more afraid of losing these rewards than providing a single piece without compensation. Even without altruists, however, "cheap pseudonyms" can provide the necessary uncertainty to make sharing rational, too.

Result 3: *In the BitTorrent game with cheap pseudonyms, sharing can be rational.*

Here, cheap pseudonyms allow peers to form groups, each acting as a single entity. Composed of multiple peers and renewed over time with new ones, these entities may persist indefinitely, mimicking the play of an infinitely repeated game between groups. This is in surprising contrast to previous work suggesting cheap pseudonyms are a negative, not useful, feature [2].

3 Future Work

In future work, we will extend our model to formally consider piece revelation strategies, preferential attachment, sharing neighborhoods and bandwidth limits. However, we do not expect our current results on rationality to be affected by incorporating these elements, as our current model is sufficiently general to simulate many of the effects of these refinements.

References

1. Feldman, M., Lai, K., Stoica, I., Chuang, J.: Robust Incentive Techniques for Peer-to-Peer Networks. In: Proceedings of the 5th ACM Conference on Electronic Commerce, New York, NY (May 2004)
2. Friedman, E., Resnick, P.: The Social Cost of Cheap Pseudonyms. Journal of Economics and Management Strategy 10(2), 173–199 (2001)
3. Immorlica, N., Lucier, B., Rogers, B.: Emergence of Cooperation in Anonymous Social Networks through Social Capital. In: Proceedings of the 11th ACM Conference on Electronic Commerce, Cambridge, MA (June 2010)
4. Kandori, M.: Social Norms and Community Enforcement. Review of Economic Studies 59, 63–80 (1992)

Hierarchical Package Bidding: Computational Complexity and Bidder Behavior

Riko Jacob, Tobias Scheffel, Georg Ziegler, and Martin Bichler

Dept. of Informatics, TU München, Munich Germany
ziegler@in.tum.de
http://dss.in.tum.de

Extended Abstract

Hierarchical package bidding (HPB) is the first combinatorial auction format used by the US Federal Communications Commission (FCC) for the sale of spectrum. This can be considered a major breakthrough after more than 14 years of discussion on the design of a combinatorial auction for the FCC. In HPB, all licenses are prepackaged in a hierarchical manner and bidders can only submit OR-bids on packages defined in this hierarchy, which leads to linear time complexity of the winner determination. A strength of HPB and a reason for the choice by the US FCC was this computational simplicity when determining the allocation and ask prices. While HPB allows more expressiveness than the Simultaneous Multiround Auction (SMR), the number of allowed package bids is restricted by the hierarchy imposed on the items by the auctioneer. Obviously, if the hierarchy does not fit the bidders' preferences, the OR bidding language of HPB can cause exposure problems as in a simultaneous auction with complementary valuations, and similar equilibrium strategies apply as in SMR. So far, the analysis of HPB is limited to a set of laboratory experiments conducted by Goeree and Holt.

We extend this analysis in two ways. First, the OR bidding language can pose a severe limitation in many applications. The auctioneer or bidders might want to use an XOR bidding language or use various other side constraints in the winner determination. Such constraints can limit the number of items a single bidder is able to win or the overall budget that a bidder will spend. Side constraints are important in many domains. Spectrum auctions, which have been the driving application for much research in this area, regularly face spectrum caps. In procurement applications, side constraints are the rule rather than an exception. We analyze the computational complexity of HPB with additional constraints and show that HPB loses its computational virtues and the winner determination problem becomes $\mathcal{NP}$-hard as soon as such constraints are present.

Second, we try to understand behavioral reasons for inefficiency of HPB. We conducted another series of lab experiments. In one set of experiments, we replicated the experimental design with global synergies introduced by Goeree and Holt. In another set of experiments, we conducted an experiment with local synergies. One prominent example of such synergies are spectrum auctions with

P. Coles et al. (Eds.): AMMA 2011, LNICST 80, pp. 36–37, 2012.

regional licenses as in the USA. The differences in efficiency and revenue in our experiments is statistically significant to the ones conducted by Goeree and Holt, but it is still small. However, we find several interesting reasons for these inefficiencies both in the global and local value model relevant to regulators and auctioneers. For example, HPB can increase the coordination problem of smaller bidders and favor the national bidder who is interested in almost all items. Reasons include jump bids which are bids above the required bid prices and activity rules employed, as well as exposure problems for small bidders in HPB. We also monitor which packages were evaluated by bidders, and when they evaluated these packages. Interestingly, we find that most bidders preselect a small number of packages and fail to evaluate new ones throughout the auction in spite of sufficient time and monetary incentives. This preselection actually accqunts for most of the inefficiency, which we find in HPB and might be an explanation for inefficiencies in combinatorial auctions in general.

Keywords: combinatorial auctions, computational complexity, bidding languages, bidder behavior.

Efficiency, Auctioneer Revenue, and Bidding Behavior in the Combinatorial Clock Auction

Martin Bichler, Pasha Shabalin, and Jürgen Wolf

Dept. of Informatics, TU München, Munich, Germany
bichler@in.tum.de
http://dss.in.tum.de

Abstract

The Combinatorial Clock Auction (CCA) is a two-phase auction format which was designed for high efficiency and to incentivize truthful bidding. The auction has recently been used for spectrum sales in a number of countries. In this study we experimentally analyze the performance of the CCA in comparison to the Simultaneous Multi-Round Auctions (SMRA), which was used for spectrum sales worldwide for more than 15 years.

The study highlights behavioral patterns, which have largely been ignored in the discussion on spectrum auction design, but are important for regulators and for the further development of combinatorial auction designs. In particular, our results indicate that in a multiband setting with thousands of possible packages efficiency and revenue of the CCA can be significantly lower than that of the SMRA.

We conducted *experiments* (i) in a base setting which closely resembles the 2.6 GHz auction setting of many countries and (ii) in a multi-band setting resembling an environment in which different spectrum bands are sold simultaneously. To account for the high levels of bidder preparation in spectrum auctions in the field and to improve the external validity of our experiments we also conducted *competitions*. Subjects in these treatments received additional information about known auction tactics and equilibrium strategies. They participated in teams of two people and were asked to prepare a bidding strategy two weeks prior to the lab session.

A main result of our experiments is that the CCA did not yield higher efficiency in the smaller base value model, but efficiency was significantly lower than in the SMRA in the multiband value model. Also, revenue was significantly lower in all treatments and sometimes items remained unsold in spite of sufficient demand. This was due to the low number of bundle bids and the CCA payment rule. In numerical experiments we show that if bidders submitted bids on all possible bundles truthfully the CCA revenue in our setting would have been similar to the revenue of SMRA in the lab.

In the CCA, bidders only submitted a small subset of all possible bundle bids. In the multiband value model only 8.33 bids out of 2,400 possible bids were submitted by bidders in the supplementary bids round, which was a major

P. Coles et al. (Eds.): AMMA 2011, LNICST 80, pp. 38–39, 2012.

reason for inefficiency and unsold items. Bidders used heuristics to select bundles, mainly based on their strength and the synergies in the value model. In the base value model, strong bidders avoided bids on small bundles, while weak bidders bid on smaller combinations as well. According to theoretical predictions bundle bids were either at or slightly below the valuation.

The observed restrictive bundle selection is more difficult to explain theoretically. The bidders had sufficient time to submit as many bundle bids as they wanted. One explanation is that neither bidders in the lab nor in the competition had fully understood the theory of core-selecting payment rules. It is also possible that bidders felt unable to submit all possible bundle bids in the multiband value model, and they bid on those that either had the highest payoff after the primary bid rounds (suggesting they had the highest chance of winning), or that had the highest synergies. Such behavior could be explained as satisficing, since bidding optimally did not seem practical.

Bidding behavior in CCA was not significantly different in competitions than in the lab. All results from the lab carry over to the competition. In SMRA treatments bidder payoff was significantly higher than in the lab indicating the successful bidder preparation.

The results need to be interpreted with the necessary care. We aimed to provide an implementation of both the CCA and the SMRA, which exactly mirrors the relevant auction rules used in the field to address external validity. We also tried to reflect the main characteristics of valuations telecoms had for spectrum in the recent spectrum sales across Europe. At the same time we limited the complexity, which bidders were exposed to in the lab. The synergies in our valuations were considerable (up to 80%), but one could also argue for even stronger synergies in some settings. Also, our experiments were conducted in a pure private values environment, while in some applications values might be affiliated.

A Framework for Automated Bundling and Pricing Using Purchase Data

Michael Benisch and Tuomas Sandholm

School of Computer Science, Carnegie Mellon University

Abstract. We present a framework for automatically suggesting high-profit bundle discounts based on historical customer purchase data. We develop several search algorithms that identify profit-maximizing prices and bundle discounts. We introduce a richer probabilistic valuation model than prior work by capturing complementarity, substitutability, and co-variance, and we provide a hybrid search technique for fitting such a model to historical shopping cart data. As new purchase data is collected, it is integrated into the valuation model, leading to an online technique that continually refines prices and bundle discounts. To our knowledge, this is the first paper to study bundle discounting using shopping cart data. We conduct computational experiments using our fitting and pricing algorithms that demonstrate several conditions under which offering discounts on bundles can benefit the seller, the buyer, and the economy as a whole. One of our main findings is that, in contrast to products typically suggested by recommender systems, the most profitable products to offer bundle discounts on appear to be those that are occasionally purchased together and often separately.

1 Introduction

Business-to-customer retail sales account for nearly four trillion dollars in the United States annually, and the percentage of this shopping done online increased three-fold between 2002 and 2007 [25]. Yet, despite the increased computational power, connectivity, and data available today, most online and brick-and-mortar retail mechanisms remain nearly identical to their centuries-old original form: item-only catalog pricing (i.e., take-it-or-leave-it offers). These are the default of B2C trade and are used by massive online retailers like Amazon, Best Buy, and Dell. However, they are fundamentally inexpressive because they do not allow sellers to offer discounts on different combinations, or *bundles*, of items.

Recently, some electronic retailers have started offering large numbers of bundle discounts (e.g., motherboards and memory at the popular computer hardware site, New Egg, and songs or albums on music sites), and brick-and-mortar retailers often offer bundle discounts on select items, such as food and drinks. Such discounts make the item-only catalog more expressive, and can be viewed as part of the general trend toward increased expressiveness in economic mechanisms. Increases in expressiveness have been shown to yield better outcomes in the design of general economic mechanisms [6], and in a number of specific domains such as sourcing auctions [23] and advertisement markets [7,28].

P. Coles et al. (Eds.): AMMA 2011, LNICST 80, pp. 40–52, 2012.

Researchers in economics, operations research, and computer science have studied issues surrounding choosing prices and bundles in various types of catalog settings for decades. However, this work has either been i) largely theoretical in nature rather than operational (e.g., [1, 3, 13, 20, 24]), ii) focused on specific types of customer survey data which are not available in many settings (e.g., [15, 17, 22]), or iii) focused on specific sub-problems (e.g., pricing information goods [4, 9, 16, 19, 29], item-only pricing [5], or unit-demand and single-minded customers [14]). Despite the ability to collect substantial amounts of data about actual customer responses to different pricing schemes, retailers in most domains are still lacking practical techniques to identify promising bundle discounts.

In this paper, we introduce an automated framework that suggests profit-maximizing prices, bundles, and discounts, the first, to our knowledge, to attempt bundle discounting using shopping cart data. Our framework uses a pricing algorithm to compute high-profit prices and a fitting algorithm to estimate a customer valuation model. As new purchase data is collected, it can be integrated into the model fitting process, leading to an online technique that continually refines prices and discounts.

In Section 4, we conduct computational experiments that test each component of our framework individually and one set that tests the framework as a whole. Our results reveal that, in contrast to the products typically suggested by recommender systems, the most profitable products to offer bundle discounts on appear to be those that are only occasionally purchased together and often separately. We also use data from a classic shopping cart generator [2] to estimate the gains in profit and surplus that can be expected by using our framework in a realistic setting. We conservatively estimate that a seller with shopping cart data like that of the generator, who already has optimally priced items, can increase profits by almost 3% and surplus by over 8% using only bundles of size two (even if he has a thousand items for sale). All of our results taken together suggest that this line of work could have material practical implications.

The setting we consider involves a seller with m different kinds of items who wishes to choose a set of prices to offer on different combinations of those items to one customer at a time. However, we generalize our framework to consider settings with more than one customer by measuring expectations for profit and revenue, which implies that item prices cannot depend on the identity of the customer. We also consider the special case where a seller can only offer discounts on bundles and must hold the item prices fixed for some exogenous reason (e.g., due to existing policies or competition). We assume the seller has a cost function that can be approximated by assigning each item a fixed cost per unit sold (in the case of digital goods, which have no marginal cost to produce, we assume the seller can estimate some form of amortized cost), and his goal is to maximize expected *profit* (revenue minus cost). The seller chooses a price catalog, $\pi(b)$, which specifies a take-it-or-leave-it price for each bundle, b, of items. In an item-priced catalog, the price of a bundle is the sum of its parts. (We will be studying richer price catalogs than that, but we still will not be pricing each bundle separately in order to keep the process tractable.) The customer has a valuation,

$v(b)$, for each bundle b and chooses to purchase the bundle that maximizes her *surplus* (valuation minus price). We make the usual assumption of free disposal (i.e., the value of a bundle is at least as much as the value of any sub-bundle). We measure expected values of revenue, seller's profit, surplus, and *efficiency* (buyer's surplus plus seller's profit).

Using this model, we can easily prove that an item-price-only catalog is arbitrarily inefficient for some valuation distributions. This follows from our recent application-independent theory that proves an upper bound on efficiency based on the expressiveness of the mechanism [6].

2 Searching for Profit Maximizing Prices

To study the impact of the item-only price catalog's inexpressiveness in practice, we first develop pricing algorithms that can determine the seller's profit-maximizing prices for a given type of catalog, cost function, and distribution over customer valuations. Each of our algorithms takes as input an estimate of the buyer's probability function, $\hat{P}$, the seller's cost function, $c(b)$, a set of priceable bundles, $\hat{B}$ (determined by the type of catalog), lower and upper bounds on the price of each bundle, $L(b)$ and $U(b)$ (also determined by the type of catalog, and can be used to ensure certain prices are fixed), and a seed price catalog, $\pi^{(0)}$ (which need not be intelligently generated). We assume that the algorithm can choose any arbitrary prices for the different bundles as long as the price of a bundle is no greater than the sum of prices for any collection of sub-bundles that contain all of its items.[1] The algorithms each call $\hat{P}$ repeatedly with candidate catalogs in order to identify the one with the highest expected profit: $\max_\pi \sum_{b \in B} \hat{P}(b|\pi) \times (\pi(b) - c(b))$.

Exhaustive Pricing (EX): For each priceable bundle, $\hat{b} \in \hat{B}$, this algorithm discretizes the space between $L(\hat{b})$ and $U(\hat{b})$ into k evenly-spaced prices and checks the expected profit of every possible mapping of prices to priceable bundles. It finds an optimal solution (subject to discretization), but is intractable with more than two items and even with two items if k is too large. For a fully expressive catalog (i.e., one where each bundle is priced separately) with m items, this algorithm calls $\hat{P}$ with k^{2^m-1} different catalogs, and $\hat{P}$ can, itself, be costly to compute. Thus, we propose this algorithm be used primarily as a tool to compare results with the other algorithms on small instances.

Hill-Climbing Pricing (HC): Starting with the seed catalog, this algorithm computes the improvement in expected profit achieved by adding or subtracting a fixed Δ from each priceable bundle, which involves $2|\hat{B}|$ calls to $\hat{P}$, in each step. It updates the catalog with the change that leads to the greatest improvement, and repeats this process until there are no more improving changes. The resulting catalog is returned, and, since the catalog is only updated when an improvement is possible, it is guaranteed to have the highest observed expected revenue.

[1] This ensures the catalog is consistent so that a customer cannot get a better price on a bundle by purchasing its components in some other combinations.

Gradient-Ascent Pricing (GA): Starting with the seed catalog, this algorithm computes the gradient, or partial derivative, of the expected profit function, which involves $|\hat{B}|$ calls to $\hat{P}$ in each step. The partial derivative, $\boldsymbol{d}(b)$, of the expected profit function with respect to a bundle, b, is estimated by measuring the change in expected profit when a fixed Δ is added to $\pi(b)$. The resulting vector of derivatives, $\boldsymbol{d}$, is normalized to sum to one, and the algorithm updates its best candidate catalog by adding $\boldsymbol{d}(b) \times \Delta$ to the price of each priceable bundle. The algorithm continues this process until no more improvements in expected profit are possible. The resulting catalog is returned, and, as with the hill-climbing algorithm, it is guaranteed to be the one with the highest expected revenue that was explored throughout the search. In our experiments, this algorithm achieved near-optimal expected revenue on most instances, while performing poorly on a few, with a relatively few number of calls to $\hat{P}$.

Pivot-Based Pricing (PVT): This algorithm generalizes hill-climbing by searching for the best adjustment to the current prices of up to k bundles at a time. For each k-or-less-sized combination of priceable bundles, β, this algorithm measures the change in expected profit from simultaneously adjusting all the prices in β. Each price can be incremented by Δ, decremented by Δ, or not changed. At each step, the algorithm tests all of those possibilities and selects the one that increases expected profit the most. The hill-climbing algorithm above is a special case of this where $k = 1$. However, for larger values of k it generalizes that algorithm to consider more complex types of price adjustments. This process involves $\binom{|\hat{B}|}{k} \times (3^k - 1)$ calls to $\hat{P}$ at each step. Even with $k = 2$, our early tests show this is the only one of the algorithms (other than the exhaustive one), that achieves optimal expected revenue on nearly every instance.

3 Estimating a Rich Customer Valuation Model

The problem of estimating a customer valuation model from historical purchase data is an essential part of our bundling framework because it allows us to use the pricing algorithms presented in the previous section in a practical setting. It is also a problem of interest in its own right, as it extends the classic *market basket analysis* problem first introduced by Agrawal *et al.* [2]. Market basket analysis is a commonly studied data mining problem that involves counting the frequencies of different bundles in a collection of customer purchase histories. Simply counting these occurrences can be challenging when there is a large set of items and each customer buys several of them at once. Almost all of the work on this problem has focused on building recommender systems that suggest products frequently purchased together. Many algorithms have been developed for finding bundles with such statistical properties, including one that was developed and patented by Google co-founder Sergey Brin and others [8]. However, as our experiments in Section 4 show, our framework predicts that the most profitable items to bundle are those with the opposite profile.

The valuation modeling problem that we consider extends the market basket analysis problem to involve predictions about what would happen to the purchase frequencies under different price catalogs. (There has been significant recent progress on inferring valuation distributions from bids or other indications of demand in a variety of applications [17,18,21,26], but that work typically used bids in auctions or survey information.)

The inputs to the two problems are essentially the same, although in the case of our valuation problem we include the price catalogs that were on offer at the time of purchase, which can provide additional information about sensitivities to price changes. The close relationship between these two problems allows us to use a classic data generator for the market basket problem in our experiments.

3.1 Deriving the Maximum Likelihood Estimate

For the valuation modeling problem, we are given a set of historical purchase observations, $D = \{\langle b_1, \pi_1\rangle, \langle b_2, \pi_2\rangle, \ldots, \langle b_n, \pi_n\rangle\}$, where each observation, i, includes a bundle that was purchased, b_i, by a distinct customer, i, and the prices of all bundles at the time, π_i. We assume that these purchases are made based on each customer's surplus-maximizing behavior with valuations drawn from an underlying valuation model. We also assume that each purchase is independent of all others since we consider each observation to be from a distinct customer. Under these assumptions, it is relatively straightforward to show that the maximum likelihood estimate (i.e., model that maximizes the likelihood of the data) for the customer valuations yields a $\hat{P}$ that matches the observed purchase frequencies as closely as possible. (Details omitted due to space constraints.)

3.2 Fitting the Valuation Model to Purchase Data

The valuation model we will fit allows for normally distributed valuations on each item, pair-wise covariance between valuations for items, as well as normally distributed terms for complementarity (or substitutability in case such a term is negative). This model significantly generalizes prior ones [10,11,24,27] by allowing for heterogeneous complementarity and substitutability between products.

Specifically, our model parameters include a mean and variance for each priceable bundle in $\hat{B}$ and covariances between individual items' valuations. While the draw, $x_{\{i\}}$, from the distribution of an item i represents that item's valuation, $v(\{i\})$, to the customer, a draw from the distribution for a bundle b of two or more items represents a complementarity bonus (or substitutability penalty if negative). The valuation for a bundle is then the sum of the draws of all the bundles (including individual items) it contains: $v(b) = \sum_{b' \subseteq b} x_{b'}$. Under this model, a customer's valuation can be thought of as a hyper-graph where each (hyper-)edge is associated with a real-valued random variable representing the valuation bonus or penalty for receiving a bundle containing the items connected by the (hyper-)edge. This allows us to model any possible distribution over valuations (without loss of generality), and can be viewed as a probabilistic generalization of the classic $k-$wise valuation model introduced by Conitzer and Sandholm for combinatorial auctions [12].

To go from a valuation model to the probability function, $\hat{P}$, we use a Monte-Carlo method to sample customers (10,000 in our experiments) according to the valuation distribution, and, for a given catalog, we simulate their surplus-maximizing purchasing behavior (taking into account that disposal is free). This simulation is relatively straightforward since items that are not connected by a complementarity or substitutability edge can be considered independently.

In order to identify the model parameters that maximize the likelihood of the observed data, we use a hybrid search technique. It begins by performing a tree search over the variance and covariance parameters. A range for each of these parameters is given as input that is discretized into a specified number of values (in our experiments we use six values per parameter). At each leaf node, a local search is performed to find the means that maximize the data likelihood given the values of the variance parameters at that leaf (see Figure 1).[2] In our experiments, we use a pivot-based search, as described in Section 2, for this step. The parameter settings resulting in the highest overall likelihood are returned, and in the case of a tie an even mixture of all the tied models is used (i.e., simulated customers are sampled from each with equal probability).

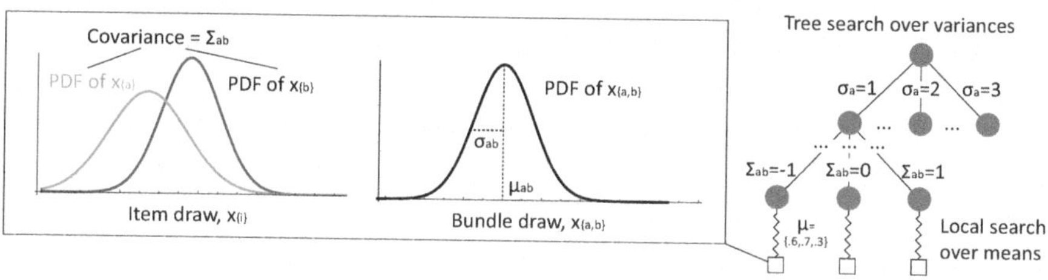

Fig. 1. Illustration of the search technique we use for estimating a customer valuation model from historical purchase data. We use a tree search over variances and a pivot-based search over means. Leaves are evaluated based on how closely the corresponding model predicts the observed data and (optionally) how closely the model's optimal profit matches the profit achieved by existing prices.

Most existing shopping cart data involve only a single catalog and do not include information about customers' surplus-maximizing behavior under alternative prices. Thus, this data tends to be under-specified for the purposes of inferring a valuation model. To address this, on such instances we utilize the existing item prices as an additional piece of information to fit our model. Specifically, among models that fit the observed purchase data (approximately), we prefer models whose profit under the optimal item-pricing for that model is close to the profit of the existing item prices under the model. Our algorithm does this test once at every leaf of the search tree (after the best model for the leaf

[2] We found empirically that choosing standard deviations with tree search and means with local search yielded better results due to the tight relationship between the appropriate means and standard deviations.

has been computed as described above). If there are still several leaves that are (approximately) as good at explaining the purchase data and the existing prices, we use an even mixture over those models (we use at most the top five models).

4 Empirical Results

We will now discuss the results from several sets of computational experiments that test our pricing and fitting algorithms and reveal some interesting economic insights that emerge as a consequence of our customer valuation model. The next two subsections focus on pricing and fitting two-item instances. The third set of results provides an estimate of the potential achievable by offering bundle discounts on pairs of items from a seller with a thousand items and realistic shopping cart data.

4.1 Results with Pricing Algorithms

The first set of experiments involves using the search techniques described in Section 2 to find high-profit prices on a generic class of instances similar to the models used in prior work [10,11,24,27]. We compare the results and performance of the pricing algorithms on symmetric two-item instances where the customer's valuation for each item is drawn from a normal distribution with mean 0.5 and standard deviation 0.5. We vary the pairwise covariance from −.25 to .25 and we vary the mean of the pair-wise complementarity (or substitutability when negative) term from −1.5 to 0.5 (the standard deviation for this term is held constant at 0.5). Each algorithm (other than the exhaustive one) uses an item-only catalog with all prices set to 0.5 as a seed and a step size $\Delta = 0.05$ to price fully expressive catalogs. The EX algorithm considers $k = 15$ different prices for each bundle and finds the optimal prices subject to this discretization. The PVT algorithm considers all possible gradients for two item instances.

The following table reports each algorithm's average fraction of the highest expected profit, efficiency, and surplus, as well as the average number of calls to $\hat{P}$ over five instances with a fully expressive catalog for each parameter setting. The best value in each column is in bold.

Algorithm	E[prof.]	E[eff.]	E[surp.]	$\hat{P}$ calls
PVT	**99.99%**	97.88%	86.92%	267.25
EX ($k = 15$)	99.24%	97.70%	88.28%	3375.00
GA	97.57%	**99.14%**	90.78%	49.61
HC	89.76%	93.73%	**96.32%**	**4.08**

Other than the unscalable exhaustive algorithm, the pivot-based algorithm is the only one to achieve optimal profit on every instance. Therefore, it is the algorithm we use in the rest of the paper for pricing. (Gradient ascent also performed well and may scale better for larger instances.)

Figure 2 shows the increase in expected profit and surplus from allowing sellers to offer profit-maximizing bundle discounts, while varying the levels of

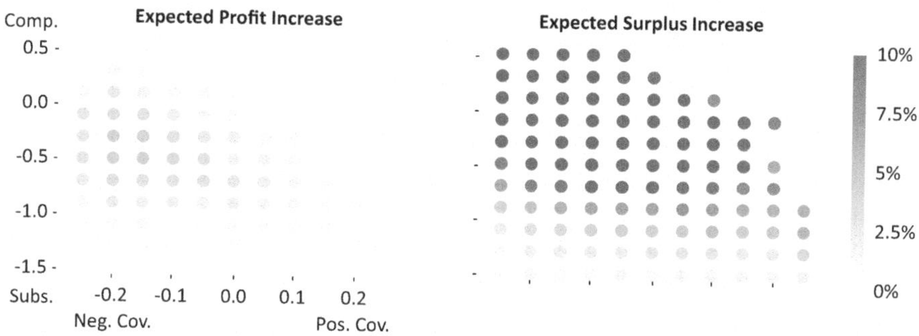

Fig. 2. The intensity of each dot is the increase in expected profit or surplus achieved by profit-maximizing bundle discounts for different levels of covariance (x-axis) and complementarity (or substitutability) (y-axis), ranging from 0% to 10%. Here, we assume the seller holds the item prices fixed at the optimal item-only catalog values to isolate the impact of bundle discounts.

covariance, complementarity, and substitutability. The values represent averages over five runs but deviate very little.

For this set of results, we assume the seller holds the item prices fixed at the optimal item-only catalog values to isolate the impact of offering bundle discounts from the potential confound of our system improving the item prices as well. We believe this also represents a practical constraint in many markets and is a policy that sellers are likely to take when first adopting the bundle discounts suggested by our framework. This has the effect of depressing the seller's expected profit gain, but it ensures that the customer surplus cannot decrease.

For the scenarios we consider, the seller's greatest predicted increase in expected profit (about 4.6%) occurs when valuations are highly negatively correlated and the items are slightly substitutable. However, too much substitutability diminishes the predicted profit benefits. Others have also identified negative correlation and substitutability as motivators for offering bundle discounts [10, 11, 27], but they did not use a rich enough valuation model to fully explore the impact of heterogeneous complementarity or substitutability. (That work also did not address the model fitting problem that must be solved to operationalize this insight.)

Unsurprisingly, due to the discount-only pricing we imposed, our results also show a large predicted increase in surplus (averaging around 9%) throughout the parameter space. Together with the seller's predicted increase in profit, this leads to substantial efficiency increases.

Another set of experiments (not shown due to space constraints) demonstrates that when our system is also free to adjust the prices of the items, additional increases in profit are possible but usually at the expense of the customer surplus. This may be desirable for the seller in the short term, but maintaining surplus can be an important long-term goal if there are competing sellers.

4.2 Results with the Fitting Algorithm

We now present experiments that use the fitting algorithm from Section 3 to find models that predict an observed set of purchase data. We allow the search algorithm to consider standard deviations between 0.5 to 3.5 at intervals of 0.5, and we focus on symmetric two-item instances where both items occur with the same frequency in the shopping cart data. (Results on asymmetric instances were similar.) These experiments test our algorithm in the ubiquitous scenario where shopping cart data is accompanied by a single item-only price catalog.

Figure 3 shows the predicted increases in expected profit and surplus achievable by a bundle discount, assuming that the individual items are optimally priced and that at those prices they have the same profit margin (the value of the profit margin does not matter). As in Figure 2, we assume the seller can only offer a discount on the existing item prices and cannot change them. (When we relax this assumption, we find additional opportunities to increase profit at the expense of customer surplus.) We consider instances where the item frequencies range from 2.5% to 40% and the *co-occurrence* percentages from 2.5% to 87.5%. We define co-occurrence as the fraction of baskets containing the less frequent item (for symmetric items either can be used) that also contain the other. We also increased our sampling frequency in an interesting area of the parameter space where item frequency is less than 15% and co-occurence is less than 20%. This is illustrated on each chart by a higher concentration of small points in the bottom left corner. (Again, the values are averaged over five runs.)

These results are consistent with those in Figure 2, since the seller's greatest predicted increase in expected profit (about 4.6%) occurs when products are occasionally bought together (co-occurrence probability less than 20%) and

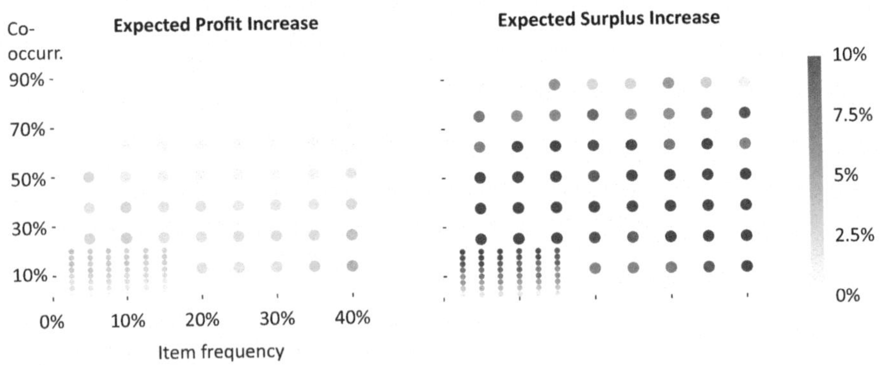

Fig. 3. The intensity of each dot represents the predicted increase in expected profit or surplus achieved by profit-maximizing bundle discounts on single-catalog instances with varying item frequencies (x-axis) and co-occurrence percentages (y-axis), ranging from 0% to 10%. As in Figure 2, we assume the seller holds the item prices fixed at the optimal item-only catalog values to isolate the impact of offering bundle discounts from the potential confound of our system improving the item pricing as well.

frequently bought separately. This set of results also predicts large increases in surplus throughout the parameter space (averaging about 9%), as in Figure 2.

Taken together, our results illustrate why new techniques are needed beyond those used for building recommender systems, which typically identify items that are commonly purchased or consumed together. When it comes to items that can be profitably bundled together at a discount, our framework suggests those with the opposite profile. Our results also explain why recommender systems are highly popular among users: a recommendation can be viewed as a small discount (in the form of time saved), and we see that even a small discount on highly co-occurring products leads to a substantial increase in surplus.

4.3 Results with a Shopping Cart Generator

Our final set of experiments estimates the potential increase in expected profit and surplus achievable by bundling products from a seller with shopping cart data like that generated by Agrawal and Srikant's classic generator [2]. We use the standard parameters in the generator: for each instance, we generate 10,000 shopping carts with 100-1,000 items (N), 100-2,000 potentially popular bundles (L) of size 2-4 (I), and an average of 2-20 purchases per customer (B). We assume the seller had optimally priced the individual items, and that those prices involved a uniform profit across all items.

Pricing all—or a huge number of—bundles is undesirable for several reasons: i) presenting complex catalogs to customers may be infeasible and/or it may confuse/burden them, ii) it is intractable in terms of computation and information, and iii) even non-overlapping bundles can interact: as one bundle is discounted, some customers might shift from buying other things to that bundle. Therefore, we only consider discounting bundles of two items, and further narrow them down as follows. We only consider item pairs priceable if the items are not directly or indirectly *related* to any other items. We consider two items related if their joint purchase frequency is more than a fixed threshold different than the product of their individual purchase frequencies (we use a threshold of 1% for these experiments). We construct a graph where the items are nodes and edges connect items that are related. Then, only connected components of size two and pairs of isolated items are considered priceable.

The profit and surplus increases for each priceable pair are then estimated using the results behind Figure 3 and a set of similar results on asymmetric instances. The increase for a given pair is estimated as the average value for the five most similar instances (based on the frequencies of the two items and the bundle). Priceable pairs are then greedily selected to actually be discounted based on their predicted profit increase. Once a pair is selected, all other pairs containing either of the selected items are removed from consideration. The following table shows the total predicted profit and surplus increases for various parameter settings of the generator (values are averaged over five instances).

N	B	L	I	E[prof.] inc.	E[surp.] inc.
1000	20	2000	4	2.80%	8.34%
1000	20	1000	4	1.10%	3.01%
100	2	200	2	0.89%	2.65%
100	2	100	2	0.15%	0.86%

For the standard parameter settings, the first row shows almost 3% profit increase using our algorithms to select pairs of items to bundle and discount. This increase in profit is accompanied by more than an 8% increase in customer surplus, and, thus, a significant efficiency increase. The table also shows that increasing the number of items and potentially popular bundles increases the benefits from our approach. This is because it leads to a sparser relatedness graph and, thus, increases the number of safely priceable items for our algorithms.

These figures are conservative because they assume that the seller had already priced the individual items optimally, only considers pairs of items where neither item is related to any other, and offers at most one bundle per item.

5 Conclusions and Future Research

We introduced a framework for automatically mining purchase data and suggesting profit-maximizing prices, bundles, and discounts. It uses a pricing algorithm to compute high-profit prices on items and some bundles, and a fitting algorithm to estimate a customer valuation model. New data can be integrated into model fitting in an online manner leading to continually refined prices and discounts.

Some obvious directions for future research include less conservative methods for selecting pricable bundles, discounting bundles of more than two items, and live experiments where the catalogs that we offer serve as demand queries about the customers' valuations that are then incorporated back into our model. These experiments could be carried out similarly to the ones described by Jedidi *et al.* [17], but would involve actual purchases by subjects rather than survey data.

There are also several assumptions made here that could be relaxed in future work. For example, we assumed that each purchase in the shopping cart data was independent, but it may be possible to develop a model that captures repeat purchases by the same customers. We also assumed that the cost of selling an item could be described by a marginal unit cost. It would be interesting to extend our work here to include considerations of non-linear cost functions (e.g., with large start-up costs) or limited-inventory items. Finally, we assumed that the true customer valuations were drawn from distributions that could be accurately fit by our valuation model. However, it would be interesting to consider the effects of mis-representing these valuations because, for example, they are drawn from a different kind of distribution than ours (e.g., lognormal rather than normal).

References

1. Adams, W.J., Yellen, J.L.: Commodity bundling and the burden of monopoly. Quarterly Journal of Economics 90(3), 475–498 (1976)
2. Agrawal, R., Srikant, R.: Fast algorithms for mining association rules. In: VLDB (1994)
3. Armstrong, M.: Multiproduct nonlinear pricing. Econometrica 64(1), 51–75 (1996)
4. Bakos, Y., Brynjolfsson, E.: Bundling information goods: Pricing, profits, and efficiency. Management Science 45(12), 1613–1630 (1999)
5. Balcan, M.-F., Blum, A., Mansour, Y.: Item pricing for revenue maximization. In: ACM EC (2008)
6. Benisch, M., Sadeh, N., Sandholm, T.: A theory of expressiveness in mechanisms. In: AAAI (2008)
7. Benisch, M., Sadeh, N., Sandholm, T.: Methodology for designing reasonably expressive mechanisms with application to ad auctions. In: IJCAI (2009)
8. Brin, S., Motwani, R., Ullman, J.D., Tsur, S.: Dynamic itemset counting and implication rules for market basket data. In: SIGMOD (1997)
9. Brooks, C.H., Durfee, E.H.: Toward automated pricing and bundling of information goods. In: Knowledge-based Electronic Markets (2000)
10. Bulut, Z., Grler, I., Sen, A.: Bundle pricing of inventories with stochastic demand. European Journal of Operational Research 197(3), 897–911 (2009)
11. Chu, C.S., Leslie, P., Sorensen, A.: Nearly optimal pricing for multiproduct firms. NBER Working Paper 13916 (2008)
12. Conitzer, V., Sandholm, T., Santi, P.: Combinatorial auctions with k-wise dependent valuations. In: AAAI (2005)
13. Dansby, R.E., Conrad, C.: Commodity bundling. The American Economic Review 74(2), 377–381 (1984)
14. Guruswami, V., Hartline, J.D., Karlin, A.R., Kempe, D., Kenyon, C., McSherry, F.: On profit-maximizing envy-free pricing. In: SODA (2005)
15. Hanson, W., Martin, R.K.: Optimal bundle pricing. Management Science 36(2), 155–174 (1990)
16. Hitt, L.M., Chen, P.-Y.: Bundling with customer self-selection: A simple approach to bundling low-marginal-cost goods. Management Science 51(10), 1481–1493 (2005)
17. Jedidi, K., Jagpal, S., Manchanda, P.: Measuring heterogeneous reservation prices for product bundles. Marketing Science 22(1), 107–130 (2003)
18. Jeziorski, P., Segal, I.: What makes them click: Empirical analysis of consumer demand for internet search advertising. Mimeo (June 2009)
19. Kephart, J.O., Brooks, C.H., Das, R., MacKie-Mason, J.K., Gazzale, R., Durfee, E.H.: Pricing information bundles in a dynamic environment. In: ACM EC (2001)
20. Preston McAfee, R., McMillan, J., Whinston, M.D.: Multiproduct monopoly, commodity bundling, and correlation of values. Quarterly Journal of Economics 104(2), 371–383 (1989)
21. Ostrovsky, M., Schwarz, M.: Reserve prices in internet advertising auctions: A field experiment. In: ACM EC (2011)
22. Rusmevichientong, P., Van Roy, B., Glynn, P.W.: A nonparametric approach to multiproduct pricing. Operations Research 54(1), 82–98 (2006)
23. Sandholm, T.: Expressive commerce and its application to sourcing: How we conducted $35 billion of generalized combinatorial auctions. AI Magazine 28(3), 45–58 (2007)

24. Schmalensee, R.: Gaussian demand and commodity bundling. The Journal of Business 57(1), S211–S230 (1984)
25. US Census Bureau. Statistical abstract of the United States, wholesale & retail trade: Online retail sales (2010), http://www.census.gov/
26. Varian, H.R.: The non-parametric approach to demand analysis. Econometrica 50, 945–974 (1982)
27. Venkatesh, R., Kamakura, W.: Optimal bundling and pricing under a monopoly: Contrasting complements and substitutes from independently valued products. Journal of Business 76(2), 211–231 (2003)
28. Walsh, W., Boutilier, C., Sandholm, T., Shields, R., Nemhauser, G., Parkes, D.: Automated channel abstraction for advertising auctions. In: AAAI (2010)
29. Wu, S.-Y., Hitt, L.M., Chen, P.-Y., Anandalingam, G.: Customized bundle pricing for information goods: A nonlinear mixed-integer programming approach. Management Science 54(3), 608–622 (2008)

Crowdsourced Bayesian Auctions
(Abstract)

Pablo Daniel Azar, Jing Chen, and Silvio Micali

MIT
32 Vassar Street, Cambridge, MA, 02139, USA
{azar,jingchen,silvio}@csail.mit.edu

Abstract. We investigate the problem of optimal mechanism design, where an auctioneer wants to sell a set of goods to buyers, in order to maximize revenue. In a Bayesian setting the buyers' valuations for the goods are drawn from a prior distribution $\mathcal{D}$, which is often assumed to be known by the seller. In this work, we focus on cases where the seller has no knowledge at all, and "the buyers know each other better than the seller knows them". In our model, $\mathcal{D}$ is not necessarily common knowledge. Instead, each player individually knows a posterior distribution associated with $\mathcal{D}$. Since the seller relies on the buyers' knowledge to help him set a price, we call these types of auctions *crowdsourced Bayesian auctions.*

For this crowdsourced Bayesian model and many environments of interest, we show that, *for arbitrary type distributions $\mathcal{D}$ (in particular, correlated ones),* it is possible to design mechanisms matching to a significant extent the performance of the optimal classical mechanisms where the seller knows $\mathcal{D}$. Our results are "existential" for a broad class of environments (including combinatorial auctions) and "constructive" for auctions of a single good.

To obtain our results, we use two techniques: (1) Proper scoring rules to elicit information from the bidders; and (2) a quantitative version of the classical Bulow-Klemperer inequality. The first technique lets us build mechanisms that guarantee good revenue, even when the players' second and higher-order beliefs about each other are wrong. The second allows us to *upper bound* the revenue of an optimal mechanism with n players by a $(1 - \frac{1}{n})$ fraction of the revenue of the optimal mechanism with $n - 1$ players. We believe that both techniques are new to Bayesian optimal auctions and of independent interest for future work.

Keywords: Optimal Auctions, Implementation Theory, Robust Mechanism Design, Proper Scoring Rules.

P. Coles et al. (Eds.): AMMA 2011, LNICST 80, p. 53, 2012.

Approximation of Large Games with Applications to Uniform Price Auctions

Aaron Bodoh-Creed

Cornell University, Dept. of Economics,
462 Uris Hall,
Ithaca, NY 14853
acreed@cornell.edu

Abstract. We provide a framework justifying the use of nonatomic limit model approximations to analyze the large market behavior of otherwise intractable game-theoretic models. We demonstrate continuity requirements on the economic primitives sufficient for the equilibrium strategies of the two models to converge as the number of participants in the large finite game approaches infinity. We apply our analysis framework to show that the equilibrium of a large interdependent values uniform price auction model where bidders have complementary preferences for multiple units can be approximated by a nonatomic exchange economy. We prove that the uniform price auction asymptotically aggregates idiosyncratic bidder information into market price and that the uniform price auction is approximately efficient with a large number of participants in the private values or single unit demand case.

Keywords: Approximate Equilibrium, Large Games, Uniform Price Auction, Rational Expectations Equilibrium. JEL Codes: C72, D44, D5.

P. Coles et al. (Eds.): AMMA 2011, LNICST 80, p. 54, 2012.

Side-Communication Yields Efficiency of Ascending Auctions: The Two-Items Case

Ron Lavi[1,⋆] and Sigal Oren[2,⋆⋆]

[1] Faculty of Industrial Engineering and Management
Technion – Israel Institute of Technology
ronlavi@ie.technion.ac.il
[2] Computer Science Department
Cornell University
sigal@cs.cornell.edu

Abstract. We analyze an ascending auction with anonymous item-prices, when there are two items that are substitutes. This popular auction format entails increased opportunities for bidders to coordinate bids, as the bidding process is longer, and since bidders see the other bids and can respond to various signaling. This has happened in many real auctions, e.g., in the Netherlands 3G Telecom Auction and in the FCC auctions in the US.

While on the face of it, such bidding behavior seems to harm economic efficiency, we show that side-communication may actually improve the social efficiency of the auction: We describe an ex-post sub-game perfect equilibrium, that uses limited side-communication, and is ex-post efficient. In contrast, without side-communication, we show that there is no ex-post equilibrium which is ex-post efficient in the ascending auction.

In the equilibrium strategy we suggest, bidders start by reporting their true demands at the first stages of the auction, and then perform a single demand reduction at a certain concrete point, determined using a single message exchanged between the bidders. We show that this limited signaling opportunity resolves the strategic problems of myopic bidding, and, quite surprisingly, improves social welfare instead of harming it.

Keywords: ascending auctions, myopic bidding, signaling, ex-post efficiency.

⋆ Supported in part by grants from the Israeli Science Foundation, the Bi-national Science Foundation, the Israeli ministry of science, and by the Google Inter-university center for Electronic Markets and Auctions.

⋆⋆ Work done while this author was at the Technion, supported by grants from the Israeli Science Foundation and the Bi-national Science Foundation.

P. Coles et al. (Eds.): AMMA 2011, LNICST 80, p. 55, 2012.

Improving Allocations through Revenue Redistribution in Auctions with Entry

Ruggiero Cavallo

Yahoo! Research,
111 West 40th Street,
New York, NY 10018
cavallo@yahoo-inc.com

Abstract. We consider a single-item private values auction setting with endogenous costly entry. It is known that for constant entry cost functions, the Vickrey auction has a unique symmetric Bayes-Nash "cutoff equilibrium" where only agents with value above some threshold enter. We extend this result to the case where cost is an arbitrary linear function of value. We then consider a class of strategyproof and no-deficit *redistribution auctions* that return Vickrey auction revenue back to participants (Bailey, 1997; Cavallo, 2006), demonstrating the existence of unique symmetric cutoff equilibria here also, with increased participation and expected allocation value (gross of costs) over the Vickrey auction. We show that when individuals' costs are purely a function of value, the equilibrium symmetric cutoff strategy achieved by the Vickrey auction is more efficient (net of costs) than any other cutoff level. But in the case where costs decrease as participation increases (e.g., when there is a search cost associated with *obscure* auctions), the Vickrey auction is inefficient and a redistribution auction dominates.

Keywords: auctions, redistribution mechanisms, entry costs, outside option, social welfare.

P. Coles et al. (Eds.): AMMA 2011, LNICST 80, p. 56, 2012.

Bubbles, Crashes and Efficiency with Double Auction Mechanisms*

Jinpeng Ma[1] and Qiongling Li[2]

[1] Department of Economics, Rutgers University, Camden, NJ 08102
jinpeng@crab.rutgers.edu
[2] Department of Mathematics, Rice University, 6100 Main St, Houston, TX 77005
ql4@rice.edu

"[S]ince equilibrium is a relationship between actions, and since the actions of one person must necessarily take place successively in time, it is obvious that the passage of time is essential to give the concept of equilibrium any meaning. This deserves mention since many economists appear to have been unable to find a place for time in equilibrium analysis and consequently have suggested that equilibrium must be conceived as timeless." F.A. Von Hayek (1937)

Abstract. We provide a quantitative boundary on the stepsizes of bid and ask of a double auction (DA) mechanism to answer two questions, when the DA mechanism is efficient and when it creates bubbles and crashes. The main result is that the ratio of the two stepsizes and their spread are the key factors for the DA mechanism to be efficient. Sentiment that leads to a swing in the spread and the ratio of the two stepsizes can result in prices to deviate from the intrinsic value equilibrium. These results are derived from a theoretical analysis of the DA mechanism built on the incremental subgradient method in Nedić and Bertsekas (2001).

Keywords: Double auction mechanism, incremental subgradient method, efficient markets hypothesis, investors' sentiment, job matching market, multiple objects.

* We thank Mark Satterthwaite, an expert on DA mechanisms, for helpful communications.

P. Coles et al. (Eds.): AMMA 2011, LNICST 80, p. 57, 2012.

From Boston to Shanghai to Deferred Acceptance: Theory and Experiments on a Family of School Choice Mechanisms

Yan Chen[1] and Onur Kesten[2]

[1] School of Information, University of Michigan, 105 South State Street, Ann Arbor, MI 48109-2112
yanchen@umich.edu

[2] Tepper School of Business, Carnegie Mellon University, PA 15213
okesten@andrew.cmu.edu

Abstract

School choice has been one of the most important and widely-debated education policies in the past two decades, with game theory playing a major role in the adoption of school choice mechanisms. Similarly, college admissions in China have gone through many local experimentations, with a rich variety of mechanisms used in various provinces.

In this paper, we investigate a family of proposal-refusal school choice mechanisms, including the Boston, Shanghai, and Deferred Acceptance (DA) mechanisms as special cases, and spanning the entire class of the Chinese parallel mechanisms. In our theoretical analysis, we first present a family of *proposal-refusal* mechanisms in which each member is characterized by some positive integer $e \in \mathbb{N}_+$ of rounds through which the proposal and refusal process continues before assignments are made permanent. More precisely, the mechanism works as follows: During rounds 1 through e, students apply to schools in order of reported preference from the most preferred to the least, and schools tentatively admit applicants up to their capacity in order of priority going from the highest to the lowest. At the end of round e students tentatively held at a school are permanently accepted into that school. The remaining students participate in a new proposal and refusal process from round $e+1$ through round $2e$. The process continues in this fashion until no student remains unassigned.

It is quite easy to see that as e increases, we go from the familiar Boston mechanism ($e = 1$) to the Chinese parallel mechanisms ($e \in [2, \infty)$) which include the Shanghai mechanism ($e = 2$), and from those to the DA ($e = \infty$). In this framework, we find that, as one moves from one extreme member of this family to the other, the experienced trade-offs are in terms of efficiency, stability and strategic immunity. Within this family, Boston is the only Pareto efficient mechanism given truthtelling, and DA is the only mechanism that is strategy-proof and stable. We provide property-based rankings of the members of this family following the works of Kesten (2006ab) and Pathak and Sonmez (2011). Specifically, we show that the members of this family can be ranked according to their immunity against strategic action. Under certain restrictions on the problem domain, any given member is *more manipulable* than a member with a higher e number. On the welfare side, a more subtle comparison result emerges. The number

P. Coles et al. (Eds.): AMMA 2011, LNICST 80, pp. 58–59, 2012.

of students receiving their reported first choices diminishes with an increasing e. As far as stability or Pareto efficiency is concerned, the ranking is ambiguous within the general class of mechanisms while the Shanghai mechanism is *more stable* than the Boston mechanism which is *more efficient* than the Shanghai mechanism.

Since the theoretical efficiency ranking in this family of mechanisms assumes truthtelling, which is a dominant strategy only under DA, it is important to assess the behavioral response to members of this family. On a broader level, as the theoretical ranking for stability or Pareto efficiency is ambiguous, empirical rankings of aggregate performance measures for these mechanisms in different controlled laboratory settings are informative for policymakers in reforming school choice or college admissions.

For these reasons, we evaluate three members of this family in two environments in the laboratory. In both environments, we find that participants are most likely to reveal their preferences truthfully under the DA mechanism, followed by the Shanghai and then Boston mechanisms. Consistent with theory, DA achieves a significantly higher proportion of stable outcomes than either Shanghai or Boston. However, the efficiency comparison is sensitive to the environment.

Our empirical findings on the manipulability and stability of the Boston mechanism compared to DA are consistent with earlier experimental work on school choice. However, we differ from previous research in that we present the first experimental evidence for the performance of the Shanghai mechanism. Lastly, our theoretical characterization of the complete set of Nash equilibria provides a benchmark for analyzing the experimental data, which reveals that stable Nash equilibrium outcomes are significantly more likely to arise than the unstable ones even when the latter Pareto dominates the former. To our knowledge, this is the first experimental evidence on equilibrium selection in school choice mechanisms.

Stability and Efficiency in the General-Priority-Based Assignment (Extended Abstract)

Aytek Erdil[1] and Taro Kumano[2]

[1] Faculty of Economics, University of Cambridge, Sidgwick Avenue, Cambridge, CB3 9DD, UK
nae25@cam.ac.uk

[2] Department of Economics, Washington University in St. Louis, One Brookings Drive, St. Louis, MO 63130, USA
tkumano@wustl.edu

Abstract. We introduce a general class of priority orders over sets, which captures both indifferences and substitutability. Our notion of substitutability ensures the existence of stable assignment. The characterization of efficient priority structures implies that there is usually a conflict between efficiency and stability. Thus we turn to the problem of finding a constrained efficient assignment, and give an algorithm which solves the problem for any priority structure that falls into our class. As an important application, gender equality or racial equality in school choice can be captured by our model, but not previous models in the literature.

Keywords: Stability, Efficiency, Generalized deferred acceptance algorithm, Substitutability, Indifferences, Stable agent improving cycle, Equal treatment of equal agents, Acyclicity.

Overview

We develop a general-priority-based assignment problem by introducing a general notion of *substitutability* on priority orders which captures indifferences and substitutability in the previous literature. Indifferences or ties are fairly common in practice, but only recently has received attention in the two-sided matching literature. See for example, Abdulkadiroğlu, Pathak and Roth (2009), Ehlers and Erdil (2010), and Erdil and Ergin (2006, 2008). On the other hand, *substitutability*, originally introduced by Kelso and Crawford (1982), is the most general class of the strict priority orders for the existence of a stable assignment. Our notion of substitutability combines these two notions in a unified framework, and we analyze *stability* and *efficiency* properties in this more general environment.

Since our priority structure involves indifferences, and is not necessarily responsive, a strict resolution of the priority orders may fail to preserve substitutability, and thus we do not straightforwardly guarantee a stable assignment. However, thanks to the monotonic property of our substitutability condition,

P. Coles et al. (Eds.): AMMA 2011, LNICST 80, pp. 60–61, 2012.

the deferred acceptance algorithm, first introduced by Gale and Shapley (1962), can be appropriately extended to our model to get a generalized deferred acceptance algorithm (GDA). Provided our notion of substitutability holds, the GDA algorithm always returns a stable assignment. As is the case for responsive priorities with ties, the set of constrained efficient allocations is not necessarily a singleton, and the arbitrary choices within the implementation of the GDA algorithm may lead to a constrained inefficient outcome.

Thus the question of finding a constrained efficient assignment remains. In the case of responsive priorities, the stable improvement cycles algorithm by Erdil and Ergin (2008) reaches a constrained efficient matching. We point out that their 'cycles of schools' are not sufficient to recover efficiency in the more general model. We introduce the *stable agent improving cycles* (SAIC), and by further assuming *equal treatment of equal agents*, we show that a stable assignment is constrained efficient if and only if it does not admit SAIC. Hence, we guarantee a constrained efficient assignment through the GDA and SAIC algorithms.

Further we ask when a constrained efficient outcome is indeed efficient. We formulate a general *acyclicity* condition, and show that the constrained efficient stable correspondence is efficient if and only if the priority structure is *acyclic*, generalizing Ergin (2002), and Ehlers and Erdil (2010). Acyclity is typically interpreted as the similarity of priorities of objects, however, this interpretation is no longer valid in our class. We observe that a priority structure may be cyclic even when every object is endowed with the same priority order.

As an application, our notion of substitutability can incorporate some notion of 'gender equality' or 'racial equality' in school choice. These notions are increasingly considered in many school choice districts, and our approach helps capture these notions in a novel way.

The Regulated Market for Kidneys in Iran

Farshad Fatemi*

Graduate School of Management and Economics,
Sharif University of Technology, Tehran, Iran
ffatemi@Sharif.edu

Abstract. We study the kidney market in Iran. The most effective treatment for end-stage renal disease is a kidney transplant. While the supply of cadaveric kidneys is limited, the debate has been focused on the effects of the existence of a free market for human organs. Economists as well as medical and legal researchers are divided over the issue. Iran has a unique kidney market which has been in place for over 20 years, frequently reporting surprising success in reducing the waiting list for kidneys. This paper demonstrates how the Iranian system works.

Keywords: Kidney Market.

1 Introduction

The most effective treatment for End-Stage Renal Disease (ESRD) is a kidney transplant (Renal Replacement Therapy: RRT). The only alternative treatment is dialysis and RRT is the only way for the patient to live without needing dialysis on a regular basis. Some researchers predict that the number of patient with ESRD reached 2 million worldwide by 2010 (Nwankwo et al., 2005). In the US, it is predicted that more than 40% of patients may die while on the waiting list (Matas, 2006). Xue et al. (2001) predict that more than 95,000 patients were on the waiting list for a kidney transplant by 2010; the figure was more than 65,000 in 2007.

There are two sources for a kidney transplant, cadaveric kidneys and kidneys from the live donors. Cadaveric kidneys can be harvested either from a brain-dead patient (whose heart is still beating) or cardiacally dead patient; the latter is considered to have a lower quality. Since a normal person can live on just one kidney, she can decide to donate one of her kidneys. The incentive to donate a kidney can be altruistic or obtaining money by selling a kidney.

In order to match a kidney from a donor with a potential recipient, their ABO and RdH blood types as well as tissues should be compatible. The ABO matching should follow the same rules that should be considered for blood transfusion, although some

* The author would like to convey thanks to Kidney Foundation especially Mr. Ghasemi and Mrs. Shaban-Kareh for providing access to the foundation data base. Dr. Alireza Heidary from the Organ Procurement Centre at the Iranian Ministry of Health also was very helpful in order to improve author's understanding of the Iranian kidney donation system.

P. Coles et al. (Eds.): AMMA 2011, LNICST 80, pp. 62–75, 2012.

programs are experimenting with ABO-incompatible transplantation (Gloor and Stegall, 2007). Regarding the tissue matching, a higher proportion of tissues matched between the donor and recipient will increase the probability of a successful transplantation.

It is well documented that RRT is cost effective as compared to dialysis. For example the UK national health system (NHS) data reveals that the annual average cost of dialysis is £30,800 while the cost of kidney transplantation is £17,000 following by a £5,000 annual cost of drugs which translates to the average net benefit of £241,000 for kidney transplantation over a period of 10 (UK Transplant, 2007).

In order to compare the cost of two alternatives for Iran (all data for 2008); the annual cost of hemodialysis for a patient is about Rials 47.0m. The cost of a transplant operation is about Rials 2.4m following by estimated Rials 40.0m annual expenditure on drugs. That means from the cost of point of view the transplant is preferred and the average benefit over a 10 year period is Rials 67.6m.

It is worth mentioning, the above calculations (both for UK and Iran) are only the direct benefit of the transplantation by reducing the treatment costs. Three other factors may also be considered in the cost-benefit analysis i) opportunity cost of the time spent to get dialysis treatment, ii) improved quality of life after receiving the transplant, and iii) risk of death during the surgery for donor which is reported to be between 0.03% and 0.06%. (Becker and Elias, 2007 and Matas et al, 2003).

It is also well known that kidneys from the live sources have a better quality as compared to the kidneys harvested from cadavers. NKUDIC (2007) contains statistics from US transplants which shows that the kidneys from live donors are more effective. While the 10 year graft survival probability for live kidneys is 54.7%, the same figure for a cadaveric kidney is only 39.2%. One issue that should be addressed here is the possibility that these data is affected by selection bias. In reality, patients are not randomly matched to kidneys. Terminally ill patients are more likely to receive a cadaveric kidney which becomes available with lower degree of compatibility. On the other hand, patients on better conditions can wait a bit longer to receive a more compatible live donation. Then the cadaveric kidneys may show a lower graft survival not only because of its own condition but also because of the condition of recipients.

Harvesting kidneys has been a major concern for health systems all around the world in the last few decades. In order to increase the kidneys available from cadavers, two different systems adopted. The most popular one is the opt-in system where people, who wish to donate their organs after their death, sign up to the scheme. The alternative system is the opt-out system which is practised in some European countries, including Spain and Austria. In this system, the donor's consent is presumed and a person needs to opt out the scheme if she does not want to donate her organs after death. One legitimate argument against this system is that presumed consent means that the state is considered the owner of the body of deceased person. Some consider this to be a problematic assumption (Becker and Elias, 2007). (For an economic comparison of the two systems see Abadie and Gay (2006)).

If the donor is a close relative or emotionally related to the recipient, live donation is legal in most of the countries around the world. The sale of organs is forbidden in all countries except in Iran, which has a regulated system for selling kidneys. However, there is evidence of the abuse of the system in many other countries. There are reports that patients from wealthy countries travel to poorer countries in order to buy a kidney (Boseley, 2006) which in some cases removed from donor's body without their knowledge (Patel, 1996).

In Iran a regulated system for kidney donation with monetary compensation was introduced in 1980s. Under this regime the donor receives a monetary compensation from the recipient and enjoys additional monetary and non-monetary bonuses from the government. The system has been criticised harshly (i.e. Harmon and Delmonico, 2006 and Zargooshi, 2001) as well as receiving some warm support (i.e. Daar, 2006 and Mahdavi-Mazdeh et al., 2008) both inside Iran and internationally. Ghods and Savaj (2006) show that in 2006, 1858 kidney transplantation took place in Iran. 13% and 12% of these transplants were harvested from cadaveric and emotionally related live sources respectively and the other 75% was from unrelated live donations (Pondrom, 2008).

There has been no discussion on how the Iranian system works by economists, while there were a lot of discussion in medical journals (for some of the most recent ones look at Ghods and Savaj (2006), Griffin (2007), and Mahdavi-Mazdeh et al. (2008)), the lack of publication in economics journals leads to misleading quotes in other researches. For example Becker and Elias (2007) mention that Iranian government opposes the cadaveric donation on religious grounds which in not true. On contrary, based on data from IRODaT (2009) the Iranian government tries hard to replace the live donation with harvesting kidneys from cadavers and the number of other cadaveric organ donations is also growing fast (Pondrom, 2008).

In this paper, we try to establish clearly how the Iranian regulated system works and find facts using the data collected from one procurement centre in Tehran. Our finding shows the average waiting time in Iranian system is around 5 months which shows a significant difference compared to average waiting time in other countries.

Following we start with a brief review on the economics literature of organ donation in section 2. In section 3 and 4, we demonstrate Iran's case and present the data collected from one of the procurement centres. Section 5 concludes and offers some policy implications.

2 Literature Review

Economists have made contribution to the organ donation literature in two fields, the kidney market and issues associated with that and also designing a mechanism to resolve the allocation and compatibility issues.

2.1 Kidney Market

Discussion on buying and selling organs or parts of human body (including blood) can be done on four grounds: medical, moral, legal, and economic grounds. Top medical experts do not agree on whether the organ market can be implemented or should be banned. Some of the most recent arguments for and against the idea can be found respectively in Reese et al. (2006) and Danovitch and Leichtman (2006).

From the medical point of view, the evidence as presented in introduction shows that live donation is efficient and cost effective. Furthermore, if it is safe to be performed on an emotionally related donor, there should be no medical concern for a kidney market on the medical grounds. Roth (2007) explains how the ethical and moral belief of majority of a society may affect the market as repugnance. Defining a property framework for the human body is one of the fresh lines of research in medical ethics. (i.e. Quigley, 2007).

The early discussion on the economics of a market for human body parts goes back to 1970 when Titmuss argued that buying and selling blood has an adverse effect on the quality of the blood (Titmuss, 1997). Titmuss compares the data from the British system (where paying for the blood was illegal) with the American system (where blood donors got paid) and argues that the latter had a lower quality of denoted blood. Titmuss points out that a monetary compensation system tends to decrease the supply (crowding effect) and acquire unhealthier blood.

Becker (2006) argues that even if Titmuss was right about the quality of the blood, the American system provides more blood per capita than British system, which means that the crowding out effect is not present. In case of kidneys, one can argue that medical developments can determine the well-being of the donor and recipient. On the other hand, since kidney transplant is a more complicated and costly procedure comparing to blood transfusion, the initial test for the donor in order to assess the quality of the kidney, as well as the donor's safety and welfare, would be more justifiable. Mellström and Johannesson (2008) ran a field experiment on the blood transfusion system in Sweden and conclude that the crowding out effect is observed only on some part of population (women) which will be eliminated if the monetary payment made to charity rather than the individual.

Cohen (1989), Epstein (1993), and Kaserman and Barnett (2002) discuss the monetary compensation for cadaveric organ donations but Becker and Elias (2007) are the first to calculate a price for live kidneys. They calculate a price of a kidney based on three components i) risk of death as a result of donation, ii) time lost during recovery, and iii) risk of reduced quality of life. They suggest a price of $15,200 for a kidney. They also point out that if the market for cadaveric kidneys established alongside the live kidney market, most kidneys will come from cadavers and live kidney prices works as a benchmark for the market equilibrium price for cadaveric kidneys.

2.2 Kidney Exchange Mechanisms

One of the main restrictions for emotionally related organ donations is the compatibility issue, where the donor's kidney cannot be transplanted for their intended recipient but it might be compatible with another patient who also has a non-compatible donor.

Roth et al. (2004) introduce a kidney exchange mechanism which efficiently and incentive compatibly, can increase the number of transplants. Modified versions of their model, in order to limit the number of simultaneous operations needed, with constraint on the maximum number of donor-patient pairs to two or three, has been developed in later papers (Roth et al., 2005; Roth et al., 2007; and Saidman et al., 2006).

3 Iran's Case[1]

In 1980s, dialysis equipments were scarce in Iran because of economic sanctions and lack of funds for imports (Nobakht and Ghahramani, 2006). As a result, nephrologists were encouraged to perform kidney transplants. At the beginning, the process relied on few cadaveric kidneys available, along with emotionally related donors. But the large number of patients on the waiting list, forced the authorities to establish a regulated market for living unrelated donations. Efforts of charities, established and managed by dialysis patients and their close relatives, helped to develop the market. It is notable that over a period of 10 year (1996-2006) the rate of cadaveric and live donation increased by 35.1% and 8.1% annually. Cadaveric transplants accounted for 1.6% of total number of transplantation in 1996, this figure reached 13.1% in 2006 (IRODaT, 2009).

There are several bodies involved in kidney procurement for patients in need of a kidney transplant in Iran, the main two players are:

Kidney Foundation of Iran (or *Dialysis and Transplant Patients Association (DATPA)*) is a charity founded by some of kidney patients and their relatives about 20 years ago. With 138 branches around the country, they help kidney patients with medical, financial, and other problems. In about 10 centres they have kidney donation offices. Their main and busiest office is located in Tehran.

Management Centre for Transplantation and Special Diseases which is part of the Ministry of Health and Medical Education and is responsible for cadaver transplant. This centre has different nationwide waiting lists for patients in need of various organs for transplantation and is the main (and only) centre involved in procurement of organs from cadavers. When cadaveric organs of a deceased patient become available, the centre allocates the organs (including kidneys) to transplant centres around the country considering different factors including distance and waiting time.

In summer 2007, there were around 1000 patients on their waiting list for kidney transplant. In 2005, from 1854 kidney donation, 243 cases were from cadavers. Religious and traditional views are a major barrier for cadaveric donations, however, in recent years the numbers of cadaveric transplants is increasing. An opt-in system is designed and some individuals, especially young educated Iranians, have shown interest in signing to the scheme. But in practice the relatives of the dead person have veto power and they can overrule the original decision made by the person herself, as it is the case in many other countries (Abadie and Gay, 2006).

The Kidney Foundation keeps waiting lists for kidney patients with different blood types in each of its procurement offices. There are eight different lists for different blood types (see Table 1). To be on the waiting list, a kidney patient should be referred

[1] The latter part of this section is based on our interviews with the Iranian kidney foundation staff, other sources and some published papers.

to foundation by his doctor. Since the foundation does not run initial tests on patients, some patients may enter the list when they are not medically ready for a transplant. This may cause unintended delays in the matching process. A patient is given priority in the waiting list, if he either is medically in an emergency situation or is a disabled soldier.

There is no centralised waiting list and each centre has its own waiting list. Patients are asked by foundation to book in their nearest centre but some patients enter several waiting lists (including the cadaveric waiting list) in order to minimise their waiting time. However, the centres coordinate with each other in case of imbalances (especially for emergency cases) of demand and supply within a particular blood type.

Medical staff including the members of the transplantation team have no role in identifying potential donors. When a donor (should age between 22 and 35) turns up to donate her kidney, she needs to provide certain documents; including a formal consent from either her spouse or her father (in case of un-married donors)[2]. After the initial official paperwork, she will be referred to a clinic for the initial medical tests. These tests determine whether the potential donor has any kidney problem as well as blood type and whether her kidney has two renal arteries[3]. If the transplantation team suspected any possible harm to the donor either now or in the future, the donation will be cancelled. The costs of these tests, which are not high, have to be paid by the donor herself. Since the cost of these tests (estimated around Rials 50k) is not significantly high, comparing to the monetary compensation, it does not seem to have an adverse effect on donors' decision.

After the donor passes the initial tests, the administrators contact the first patient in the same waiting list as the donor's blood type. In this stage the staff also match the physical build of the donor and the patient or at least make sure that they are not extremely different. Matches cross different blood types are rare.

If the patient who is on the top of the waiting list is not ready for the transplant, the next patient will be called, and so on, until a ready patient will be found. Then a meeting between the two parties is arranged (they are provided with a private area within the foundation building if they want to reach a private agreement) and they will be sent for tissue tests. If the tissue test gives the favourable result (According to administrators of the foundation less than 10% of the tests have a positive cross-match which effectively rules out transplantation.), a contract between the patient and the donor will be signed and they will be provided with a list of the transplant centres and doctors who perform surgery. When the patient and the donor are referred to transplant centre, a cheque from the patient will be kept at the centre to be paid to the donor after the transplant takes place. The guide price has been 25m Rials (≈ \$2660) until March 2007 for 3 years and then raised to 30m Rials (≈ \$3190).[4] This decision has been made because the foundation was worried of a decreasing trend in number of donors. In some cases, the recipient will agree to make an additional payment to the

---

[2] In Absence of next of kin, to make sure the donor is aware of her action and its consequences, she will be referred to a chartered psychologist at the coroner's office for a psychic test.

[3] Most patients are not happy to have a kidney transplant from this type, since it reduces the chance of successful transplant.

[4] The exchange rate for 20 Feb 2008: \$1 ≡ Rials 9410.

donor outside the system; it is not certain how common this practice is, but according to the foundation staff the amount of this payment is not usually big and is thought to be about 5m to 10m Rials (≈ $530 to $1060). The recipient also pays for the cost of tests, two operations, after surgery cares, and other associated costs. Insurance companies cover the medical costs of the transplant and the operations are also performed free of charge in the state-owned hospitals.

In addition, the government pays a monetary gift to the donor (currently, Rials 10m), as well as automatic provision of one year free health insurance[5], and the opportunity to attend the annual appreciation event dedicated to donors. The Charity Foundation for Special Diseases also provides the donors with a free annual medical test and high level of support, in case that the donor develops kidney problems in the future, regardless of whether this is due to the transplant or not.

Emotionally related donors also enjoy these monetary and non-monetary bonuses as well as exemption of paying hospital costs, and it gives them a good incentive to register at the foundation offices. The minimum payment of Rials 45m is around 2 years of minimum wage and equivalent to around 6 times of average monthly cost of living for a family of four in urban areas reported by the Iranian Central Bank.

The minimum current payment (45m Rials) by using PPP exchange rate is equivalent to $14,000 which is interestingly close to Becker and Elias (2007) suggestion for the market value of a kidney at $15,200 for the US. In 1980s when the sale of kidney was legal in India, donors were paid $1,603. After making this illegal in 1990s the average payment dropped to $975 (Goyal et al., 2002). Based on this paper and other researches, Becker and Elias (2007) estimate that the equivalent cost of a kidney in Indian market to US dollar in 2005 is in region of $17,078 to $17,665.

Advertising for kidney donors is banned but informal advertisement is observed and some patients manage to find donors using informal channels in order to avoid the waiting list. The foundation handle these cases with due care and such cases need to be reviewed by the foundation managing director.

In order to prevent international kidney trade, the donor and recipient are required to have the same nationality. That means an Afghan patient, who is referred to the foundation, should wait until an Afghan donor with appropriate characteristics turns up. Another issue can be Iranian nationals residing abroad and travel to Iran to buy a kidney, which is allowed under current legislation. Ghods and Savaj (2006) present the data that shows the scale of transplant tourism is very small in Iran.

By the foundation's procedure to keep 8 different waiting lists, if one assumes that the blood type distribution is the same between patients and potential donors then the waiting time would be fairly similar for all waiting lists, furthermore there will be no significant social benefit in matching between blood groups.

One of the concerns about employing the Iranian system would be the possible welfare effect on the minorities because of the different pattern of the blood type distribution in their blood types. One line of fresh research in Iranian system would be

---

5 Nobakht and Ghahramani (2006) claim that the donors are provided with a free life-long insurance which is in contrary with our findings, after interviews with the foundation staff.

to investigate the proportion of different ethnic and race minorities in the pool of kidney donors and recipients.

There are two major papers which address the donors' satisfaction issue. Zargooshi (2001) surveys 300 of kidney donors. They donated between 6 to 132 months prior to being interviewed. He finds that the majority of donors either did not receive or did not attend follow-up visits and many of them regretted their original decision. On contrary, Malakoutian et al. (2007) report a 91% satisfaction between living kidney donors. However, the latter survey interviewed the donors at the point of discharge from hospital.

4 Data

Our data contains 598 transplantations recorded in Tehran office of the Kidney Foundation between April 2006 and December 2008. Of these, 549 were live kidney donations of which 539 were traded kidney and 10 emotionally related donations. The remaining 49 transplantations took place with a cadaveric donation. In theory, the waiting lists for live and cadaveric kidneys run independently; and the coverage of our data from cadaveric transplant is not complete. As a result, our data shows a 8.2% share for the cadaveric transplantations, which is lower than around 13% based on official sources. It is clear that our finding can not be a good image of what is happening in terms of emotionally related donations, because of the small number of this type of donations in our sample. Table 1 shows the ABO and RhD blood types distribution of recipients.

Table 1. The ABO and RhD blood types distribution of recipients

	Blood Type									
	O+	A+	B+	AB+	O-	A-	B-	AB-	**Total**	
Traded	150	165	110	38	27	34	10	5	**539**	90.1%
Non-Traded	3	2	2	1	0	1	1	0	**10**	1.7%
Cadaver	15	15	11	2	1	3	2	0	**49**	8.2%
Total	**168**	**182**	**123**	**41**	**28**	**38**	**13**	**5**	**598**	

In Order to check whether the traded kidneys are biased in favour of AB blood type and are disadvantageous for O type, Table 2 demonstrates the ABO blood type distribution of recipients. Although the share of AB recipients is higher in traded cases but there is no significant difference for the share of O recipients in traded and cadaveric cases.

Table 2. The ABO blood type distribution of recipients

	Blood Type				
	O	A	B	AB	**Total**
Traded	32.8%	36.9%	22.3%	8.0%	**100%**
Non – Traded	30.0%	30.0%	30.0%	10.0%	**100%**
Cadaver	32.7%	36.7%	26.5%	4.1%	**100%**
Total	**32.8%**	**36.8%**	**22.7%**	**7.7%**	**100%**

Another concern could be discriminating against women in receiving kidneys. Traditionally in Iran, men are considered the breadwinner of the family. Despite the recent changes in the sex pattern of labour force, it is still biased in favour of male workers. Since in this view, the economic value of a man is considered to be higher, one consequence in our argument can be a higher likelihood for a male patient receiving a kidney from traded sector. Our data shows 64.9% and 67.3% of male recipient for traded and cadaver kidneys respectively which do not support such a negative effect on female patients.

On the other hand, the donors are mostly men (446 in 539 cases of traded kidneys). This can be because of the two facts. Firstly, the ages between 22 and 35; when the donation is accepted; is the fertility age; and women are less likely to be considered as potential donors. Secondly, since men are considered as the family breadwinner, it is likely that they sell their kidneys to overcome financial difficulties. Female donors count for around 18% of traded kidneys in our data; it is in contrary with the Indian case where 71% of the sold kidneys were from female donors (Goyal et al. 2002).

Table 3 demonstrates age distribution of recipients and donors of traded kidneys. It shows that 10.9% of the recipients are under the age of 20. The high number of transplants needed for relatively young patients (42.9% under the age of 40 and 65.3% under the age of 50), shows the economic and social value of these transplants. Although the foundation's policy is to limit the donors' age to 35, 10.4% of the donors are older than 35.

Table 3. Age distribution of recipients and donors

Age	Recipients		Donors	
5 – 19	59	10.9%		
20 – 24	36	6.7%	148	27.5%
25 – 29	50	9.3%	216	40.1%
30 – 34	42	7.8%	119	22.1%
35 – 39	44	8.2%	51	9.5%
40 – 44	59	10.9%	5	0.9%
45 – 59	185	34.4%		
60 – 74	63	11.7%		
75 – 79	1	0.2%		
Total	**539**	**100.0%**	**539**	**100.0%**

The joint blood type distribution of recipients and donors can be seen in Table 4. On average 94.8% of kidneys are matched to an exact blood type. In total 28 cases out of 539 are matches between different blood types. The reason behind this can be emergency cases, matches found by patients themselves out of the formal system, and especial cases (like children recipients when the size of kidney plays an important rule).

Table 4. Joint ABO and RdH frequency of transplants for recipients and donors

		Donor								Total	
		O+	A+	B+	AB+	O-	A-	B-	AB-		
Recipient	O+	149				1				**150**	27.8%
	A+	2	163							**165**	30.6%
	B+	4		104				2		**110**	20.4%
	AB+		1	1	36					**38**	7.1%
	O-	7				20				**27**	5.0%
	A-	1	4			3	26			**34**	6.3%
	B-							10		**10**	1.9%
	AB-				2				3	**5**	0.9%
Total		**163**	**168**	**105**	**38**	**24**	**26**	**12**	**3**	**539**	100.0%
		30.2%	31.2%	19.5%	7.1%	4.5%	4.8%	2.2%	0.6%	100.0%	
Own type		91.4%	97.0%	99.0%	94.7%	83.3%	100.0%	83.3%	100.0%	94.8%	

The average waiting time for patients who receive a live kidney is 149 days (Table 5). By waiting time, we mean the time gap between signing into the waiting list and the operation date. This includes the time needed for the tests and preparation before the transplant when a match initially introduced.

Assuming a similar distribution in donors and recipients population over the blood types, waiting time is expected to be the same for all waiting lists. However, the waiting time for a given waiting list is going to be affected by the following: i) Not enough donors from that blood type turn up comparing to other blood types; it can be serious when one blood type is rare; like AB- for the Iranian population; ii) When kidneys from a blood group is offered to other blood groups. In our data, type O+ recipient is likely to be slightly affected by this, as the average waiting time for them is 171 days and 8.6% of this type of kidney is allocated to other blood groups. iii) When a patient enters before he is medically ready for the transplant; we cannot check for this in our data. And iv) When a mismatch arises in testing procedure which means a 2-4 weeks is added to waiting time of the next recipient of this kidney. However, we can assume the last two have a similar effect on all waiting lists.

Table 5. Average waiting time for recipients based on the blood type of both parties

		Donor								
		O+	A+	B+	AB+	O-	A-	B-	AB-	**Average**
Recipient	O+	169				461				**171**
	A+	110	138							**137**
	B+	85		138				214		**138**
	AB+		104	32	128					**125**
	O-	163				117				**129**
	A-	92	205			249	177			**184**
	B-							124		**124**
	AB-				218				144	**174**
Average		**165**	**139**	**137**	**133**	**148**	**177**	**139**	**144**	**149**

The guideline price increased by 20% on March 2007. But our data shows no significant change in the waiting time or the number of donation. It could because of two reasons; firstly this increase has almost no significant effect in real term because of inflation (The inflation rate for 2006-07 is 18.4%). Secondly, the price that actually paid in each case can be different from this benchmark by two parties' negotiation process and it can also make that increase less significant.

Considering all of the mentioned factors, having a waiting list of around 5 months in Iranian system comparing to more than 3 years for some other countries seems a significant achievement.

5 Summary and Conclusion

In this paper, our focus was not on the moral and ethical issues surrounding the discussions. The effect of the Iranian system on reducing the waiting time for patients is significant, which based on our data is around 5 months. One should be careful in advising to ban the sale, the alternative solution practiced in other developing countries, e.g. black market for organs, might have dramatic consequences. This may result lower standards on medical conditions, as well as leaving the donors who can be vulnerable without any official support.

Following, we list the possible policy considerations:

- Since the donors might be subject to exploitation because of their social status; it needs to be guaranteed that they make an informed decision and are aware of all risks attached to their decision.
- After donation networks needs to be strengthened in order to make sure the donors receive the best support possible.
- Considering the Iranian population structure, it is expected that the demand might rise for kidneys in coming years and decades. Then, more efforts need to be put on other sources of kidneys. Cadaveric kidneys can be utilised more effectively. Unlike some developed countries, Iran faces no social barrier in new frontiers in

medical research, e.g. cloning. Investing in this area may help to eliminate the demand for live donation in the future.

- A national waiting list can reduce the waiting time as well as improving pre- and post- surgery support for both donors and recipients.
- It can be shown by a simple two separated market when the supply can be transferred from one market to the other that allowing intra-trade may improve the welfare in one market and reduce it in the other. The Iranian model restricts the social welfare by not allowing trade between different markets; however, it might be justifiable with higher expected rate of success when both donor and recipient have the same blood type.

Future research can focus on two different lines; first, a behavioural experiment on whether people who decide to donate their kidney for money, have significantly different economic attitudes; second if one can get the data on actual money exchanged hands in the donation an analysis of demand and supply for kidneys can give us a better insight into this market.

References

Abadie, A., Gay, S.R.: The Impact of Presumed Consent Legislation on Cadaveric Organ Donation: A Cross-Country Study. Journal of Health Economics 25(4), 599–620 (2006)

Becker, G.S., Elıas, J.J.: Introducing Incentives in the Market for Live and Cadaveric Organ Donations. Journal of Economic Perspectives 21(3), 3–24 (2007)

Becker, G.S.: Should the Purchase and Sale of Organs for Transplant Surgery Be Permitted? (2006), http://www.becker-posner-blog.com/

Boseley, S.: UK transplant patients go to China for organs from executed prisoners. The Guardian (April 20, 2006)

Cohen, L.R.: Increasing the Supply of Transplant Organs: The Virtues of a Futures Market. George Washington Law Review 58(1), 1–51 (1989)

Daar, A.S.: The Case for a Regulated System of Living Kidney Sales. Nature Clinical Practice Nephrology 2(11), 600–601 (2006)

Danovitch, G.M., Leichtman, A.B.: Kidney Vending: The "Trojan Horse" of Organ Transplantation. Clinical Journal of American Society of Nephrology 1(6), 1133–1135 (2006)

Epstein, R.: Organ Transplants: Is Relying on Altruism Costing Lives? American Enterprise 4(6), 50–57 (1993)

Ghods, A.J., Savaj, S.: Iranian Model of Paid and Regulated Living-Unrelated Kidney Donation. Clinical Journal of American Society of Nephrology 1(6), 1136–1145 (2006)

Gloor, J., Stegall, M.: ABO incompatible kidney transplantation. Current Opinion in Nephrology and Hypertension 16(6), 529–534 (2007)

Goyal, M., Mehta, R.L., Schneiderman, L.J., Sehgal, A.R.: Economic and Health Consequences of Selling a Kidney in India. Journal of the American Medical Association 288(13), 1589–1593 (2002)

Griffin, A.: Kidneys on Demand: Did Iran Eliminate Its Waiting List for Kidney Transplants? And if so, Where are the Kidneys Coming from? British Medical Journal 334(7592), 502–505 (2007)

Harmon, W., Delmonico, F.: Payment for Kidneys: A Government-Regulated System Is Not Ethically Achievable? Clinical Journal of American Society of Nephrology 1(6), 1146–1147 (2006)

Hausman, J.A., Leonard, G.K.: Estimation of patent licensing value using a flexible demand specification. Journal of Econometrics 139(2), 242–258 (2007)

IRODaT (International Registry of Organ Donation and Transplantation) (2009), http://www.tpm.org

Kaserman, D.L., Barnett, A.H.: The US Organ Procurement System: A Prescription for Reform. AEI Press, Washington DC. US (2002)

Mahdavi-Mazdeh, M., Heidary-Rouchi, A., Aghighi, M., Rajolani, H.: Organ and Tissue Transplantation in Iran. Saudi Journal of Kidney Diseases and Transplantation 19(1), 127–131 (2008)

Malakoutian, T., Hakemi, M.S., Nassiri, A.A., Rambod, M., Nobakht, A.H., Broumand, B., Fazel, I.: Socioeconomic Status of Iranian Living Unrelated Kidney Donors: A Multicenter Study. Transplantation Proceedings 39(4), 824–825 (2007)

Matas, A.J.: Why We Should Develop a Regulated System of Kidney Sales: A Call for Action! Clinical Journal of American Society of Nephrology 1(6), 1129–1132 (2006)

Matas, A.J., Bartlett, S.T., Leichtman, A.B., Delmonico, F.L.: Morbidity and mortality after living kidney donation, 1999–2001: survey of United States transplant centers. American Journal of Transplant. 3, 830–834 (2003)

Mellström, C., Johannesson, M.: Crowding Out in Blood Donation: Was Titmuss Right? Journal of The European Economic Association 6(4), 845–863 (2008)

NKUDIC (National Kidney and Urologic Diseases Information Clearinghouse): Kidney and Urologic Diseases; Statistics for the United States. U.S. Department of Health and Human Services (2007)

Nobakht, H.A., Ghahramani, N.: Living Unrelated Kidney Donor Transplantation in Iran. Nature Clinical Practice Nephrology 2(12) (2006)

Nwankwo, E., Bello, A.K., El Nahas, A.M.: Chronic Kidney Disease: Stemming the Global Tide. American Journal of Kidney Diseases 45(1), 201–208 (2005)

Patel, T.: India Kidney Trade. TED Case Studies 5(1) (1996)

Pondrom, S.: Is the Iranian Paid Donation Model a Success ...or Not? American Journal of Transplantation 8(3), 477–478 (2008)

Quigley, M.: Property and the body: Applying Honoré. Journal of Medical Ethics 33(11), 631–634 (2007)

Reese, P.P., Caplan, A.L., Kesselheim, A.S., Bloom, R.D.: Creating a Medical, Ethical, and Legal Framework for Complex Living Kidney Donors. Clinical Journal of American Society of Nephrology 1(6), 1148–1153 (2006)

Roth, A.E.: Repugnance as a Constraint on Markets. Journal of Economic Perspectives 21(3), 37–58 (2007)

Roth, A.E., Sönmez, T., Ünver, M.U.: Kidney Exchange. Quarterly Journal of Economics 119(2), 457–488 (2004)

Roth, A.E., Sönmez, T., Ünver, M.U.: Pairwise Kidney Exchange. Journal of Economic Theory 125(2), 151–188 (2005)

Roth, A.E., Sönmez, T., Ünver, M.U.: Efficient Kidney Exchange: Coincidence of Wants in Markets with Compatibility-Based Preferences. American Economic Review 97(3), 828–851 (2007)

Saidman, S.L., Roth, A.E., Sönmez, T., Ünver, M.U., Delmonico, F.L.: Increasing the Opportunity of Live Kidney Donation By Matching for Two and Three Way Exchanges. Transplantation 81(5), 773–782 (2006)

Titmuss, R.: The Gift Relationship: from Human Blood to Social Policy. Allen and Unwin, London (1970); Oakley, A., Aston, J. (eds): The Gift Relationship, New chapters added. LSE Books, London (1997)

UK Transplant: UK Transplant Fact Sheet 4: the Cost-Effectiveness of Transplantation (2007), http://www.uktransplant.org.uk/ukt/newsroom/fact_sheets/cost_effectiveness_of_transplantation.jsp

Xue, J.L., Ma, J.Z., Louis, T.A., Collins, A.J.: Forecast of the Number of Patients with End-Stage Renal Disease in the United States to the Year 2010. Journal of American Society of Nephrology 12(12), 2753–2758 (2001)

Zargooshi, J.: Quality of Life of Iranian Kidney "Donors". Journal of Urology 166(5), 1790–1799 (2001)

Solids - A Combinatorial Auction for a Housing Corporation

Dries R. Goossens[1], Sander Onderstal[2], and Frits C.R. Spieksma[3]

[1] PostDoc researcher for Research Foundation - Flanders
Center for Operations Research and Business Statistics (ORSTAT),
Faculty of Business and Economics, KU Leuven, Belgium
dries.goossens@econ.kuleuven.be
[2] University of Amsterdam, The Netherlands
[3] Center for Operations Research and Business Statistics (ORSTAT),
Faculty of Business and Economics, KU Leuven, Belgium

Abstract. On May 7, 2011, over one hundred bidders took part in a combinatorial auction for housing space in a newly erected building in Amsterdam (the Netherlands). This paper describes the development of this auction. We sketch our collaboration with the housing corporation that resulted in design choices with respect to first/second price, feedback, number of rounds, and tractability of the combinatorial auction. Furthermore, the winner determination problem is complicated by various municipal and building regulations that the allocation needs to satisfy. We show how these regulations can be included in an integer program that is used to solve the winner determination problem. Finally, computational experiments illustrate the tractability of this model.

Keywords: combinatorial auction, housing, auction design, integer program.

1 Introduction

To raise a building that will last for at least 200 years, with its tenants deciding how to use the building. The Dutch housing corporation Stadgenoot aims to make this happen in Amsterdam with so-called "solids". A solid is a sustainable building without predefined purpose, that offers a lot of flexibility. The Solids concept is inspired by historic canalside mansions in Amsterdam [2]. The main idea is that it is up to the tenants to decide on the use, the size, the configuration and even the rent of the space in the solid that they occupy. Stadgenoot sees solids as highly suitable for a large variety of tenants: for (large) families, with everyone getting their own area, all linked up to a shared family room and kitchen, for entrepreneurs who are looking for a living area with a work space, for students, restaurants, etc. It is Stadgenoot's position that solids should be open for everyone, including people with a small budget.

Solid spaces are delivered as shells. This means that within the building, there are walls between the solid spaces, and each solid space has access to a shaft with

P. Coles et al. (Eds.): AMMA 2011, LNICST 80, pp. 76–87, 2012.

ventilation provision, drainage, electricity, etc. However, within a solid space, it is the tenant who decides where to place partition walls, interior doors, etc. This enables the rented space to be used for a whole range of purposes: living, working, culture, or a combination of all these. Stadgenoot remains the owner of the shell; the tenants rent the solid space, and own the interior. If a tenant leaves, he can sell the interior to a next tenant (throughout this paper, he can be replaced by she, and his by her). Over time, solid spaces can grow (when they are merged with another solid space) or shrink (when they are split up), and be used in very different ways.

Apart from the functional freedom, solids are also novel for the way they will be allocated to tenants, namely using a combinatorial auction. As far as we are aware, this is the first time that a combinatorial auction will be used for allocating real estate. Indeed, traditionally, residences, business accommodations, shops, and offices are treated as different markets, each with their own rules and target groups. Typically, the characteristics of the property are announced, owners set a price and wait for an interested buyer to show up. For solids, however, a combinatorial auction seems particularly suited due to the varying preferences of the different tenants. For instance, some tenants may want sun in the evening (west side), some want a small but practical working space on the first floor, others want spacious room on the top floor. A combinatorial auction is an ideal way to take all these preferences (including different budgets) into account.

The construction of the first solid, called Solid 11, was finished half 2011. Two other solids are already under construction, and there are plans to build many others in Amsterdam, for up to 50,000 square meters in total. We refer to [1] for more information about the various solids, written from the perspective of an interested tenant. In this paper, we will focus on Solid 11, located in the west part of Amsterdam, and featuring a spacious roof terrace with a splendid view over the city. In fact, on May 7, 2011, Solid 11 was allocated to bidders using the combinatorial auction described in this contribution (see [1]). On that day, over one hundred bidders participated in the auction. We, the authors of this paper, were involved in the design of this auction. We report here which considerations were taken into account, and which precise auction rules were selected. We also discuss how we determine which bidders wins which part of the solid.

The next section is devoted to the problem description. Section 3 shows how we decided upon the auction design. The winner determination problem is formally discussed in section 4, followed by computational experiments in section 5. The paper ends with conclusions in section 6.

2 Problem Description

In essence, the problem is to develop a combinatorial auction for Stadgenoot to allocate their solids to interested bidders. A combinatorial auction is an auction where multiple items are auctioned simultaneously, and bidders are allowed to bid on one or more subsets of the items. To this end, the solid is divided into 125 lots (i.e. items), distributed over 7 floors. Any interested tenant can specify solid

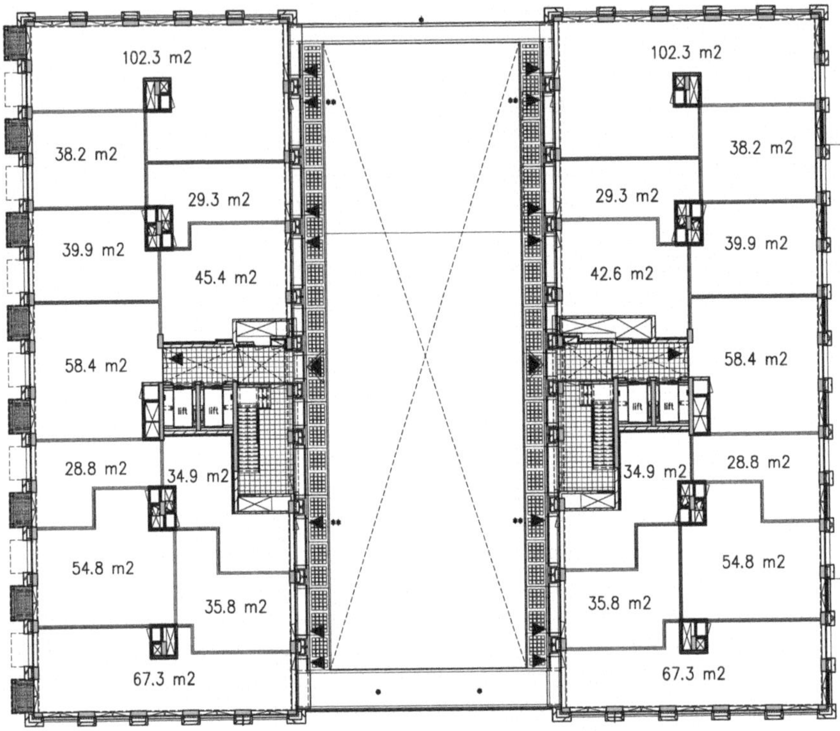

Fig. 1. Allotment of the second floor of Solid 11

spaces (i.e. sets of lots), together with the price he is willing to pay as a monthly rent (paying rent by month is the usual setting in the Netherlands). Figure 1 shows how the second floor of Solid 11 is divided into 22 lots; the surface area of each lot is given. By choosing an appropriate set of lots (on a bidder's favorite floor, according to a preferred orientation), a bidder specifies the resulting solid space.

A combinatorial auction is useful in this case, because typically, bidders are interested in multiple lots, and may value some sets of lots higher than the sum of the values of the lots individually. Moreover, some lots have no value at all as a single item, because e.g. they are not directly accessible from the hallway. These so-called complementarity effects may be bidder-specific, since bidders have different needs and preferences with respect to the space they want to rent. A combinatorial auction is an excellent way to make use of these synergies. Indeed, it allows a bidder to express his preferences to a greater extent than for individual items only, and it allows Stadgenoot to collect a higher total rent. For a thorough discussion on combinatorial auctions, integrating contributions on many interesting aspects from both theory and practice, we refer to the book edited by Cramton et al. [3].

Stadgenoot distinguishes three types of bidders: residential, commercial, and social bidders. The first group consists of people who plan to live in the solid, the second group plans to open a business in the solid, and people with a low income make up the third group. In a combinatorial auction in its most general form, bidders can bid whatever amount they please on any subset of the items. In the solid auction, this is not the case. First of all, there is a lower bound of 6 euro per square meter on each bid. Stadgenoot rather sees the space empty, than renting it out for less than 6 euro per square meter. Furthermore, for social bidders, Dutch law imposes an upper bound on the monthly rent of 650 euro. Second, and most important, is that bidders cannot bid on just any subset of lots. A valid subset consists solely of adjacent lots (on a single floor). Figure 1 shows that floors above ground level consist of two wings, separated by an open space in the middle, such that lots in different wings are not adjacent. On the ground level, the wings are connected by additional lots, and consequently, this is the only place where valid solid spaces spanning both wings are possible. A bid on a set of lots on different floors is not allowed. Furthermore, a valid subset of lots needs to have access to at least one of 10 available utility shafts, have enough doors to the central hallway (denoted by black triangles in Figure 1), and have enough incidence of light through the windows. For the remainder of this paper, when we use the term "solid space", we mean in fact sets of lots satisfying these requirements. About 1000 solid spaces can be formed with the 125 lots available in Solid 11. There are also limits to the surface area of the solid space, depending on the type of bidder. Residential bidders should bid on solid spaces of at least 90 square meters; commercial bidders can only go for solid spaces larger than 60 square meters, but there is a limit of 180 square meters for restaurants. Social bidders can only bid on solid spaces with a surface area of at most 70 square meters.

Apart from restrictions on the sets of lots bidders can bid on, there are also constraints on the allocation itself, based mainly on municipal and building regulations. The allocation should reserve at least 14.25% of the surface to social bidders. Residential bidders should get at least 25.75% of the surface, whereas commercial bidders should be allocated at least 30% of the surface. At most 3 restaurants are allowed in the solid. Furthermore, each floor above ground level has a rescue capacity per wing that should not be exceeded. The rescue capacity needed for each bid depends on the surface area of the solid space and the function that the bidder has in mind. Each solid space has to get its entire utilities through a single shaft, however, if a solid space contains multiple shafts, the choice is open which one to use. This choice can be made by the bidder, but can also be left open for the auctioneer in case the bidder has no preference. Each shaft goes from the ground level to the top floor, and has a ventilation capacity. Each bid has some ventilation requirement (again depending on the surface area and the intended function), and bids can only be allocated insofar the ventilation capacity of the shaft that they use is not exceeded. Each shaft offers a standard gas and electricity connection, however, some bidders require a connection with a higher capacity. Therefore, some shafts offer a high-capacity gas connection, on at most two floors, and a high current electricity connection on at most one floor. Bids with special requirements for gas and/or electricity can only be accepted insofar they can be accommodated by the shafts.

A concern of Stadgenoot is the possibility that a lack of bidders, or unpopular lots, would cause some parts of the solid to be unoccupied after the auction. In that case, the unoccupied lots should also form one or more solid spaces, such that they can still be rented out in the future. If there are not enough social bidders, there should be enough unoccupied solid spaces with a surface area less than 70 meters, such that when they are rented to social bidders later, the required 14.25% of the total surface of the solid can still be met. Furthermore, an unoccupied solid space should have at least one shaft where enough ventilation capacity is available to support basic use. Similarly, in order to make sure that a vacant solid space can be occupied later, sufficient rescue capacity should be reserved on its wing and floor.

For Stadgenoot, the goal is to maximize the total rent, taking into account the constraints mentioned above. Given the novelty of this way of allocating real estate, Stadgenoot wants the auction to be perceived as fair by the participants. Therefore, it is important that the auction is as transparent as possible, and accessible to everyone. Indeed, Stadgenoot expects and encourages many interested bidders with few or no experience in auctions, let alone combinatorial auctions. Therefore, the auction rules should be as simple as possible, and the bidding process should be user-friendly. Consequently, Stadgenoot decided that each bid should remain valid until the end of the auction: withdrawing or lowering bids is not allowed. Finally, each bidder can win at most one solid space, and the auction will be completed within a single day.

3 Auction Design

Every auction needs a set of rules that determines the course of the auction, the actions the bidders can take, and the feedback they will get. The design of the solid auction is based on a series of experiments with human bidders and a computer simulation [7], in which we studied the effect of various design settings on the following evaluation criteria:

- Total revenue
- Total surface of unallocated space
- Efficiency
- Auction stability
- User-friendliness
- Tractability.

The total revenue simply corresponds to the total monthly rent for Stadgenoot. Even though the total surface of unallocated space is related to the total revenue, Stadgenoot wants to avoid empty spaces, especially in the first of a series of solids that are to be auctioned. Efficiency refers to the total bidder value generated by the auction, relative to the highest possible value. This is an important measure, since an inefficient outcome implies that non-winning bidders have good reason to complain: at the final auction prices they are still willing to buy from winning bidders. The stability of the auction refers to how stable the winning allocation is over the rounds of the auction. Bidders might distrust the auction outcome if the set of winning bidders changes substantially from one round to the next, particularly in the final rounds of the auction. Indeed, we have to keep in mind that the vast majority of the bidders are not familiar with auctions. Therefore, user-friendliness of the auction and the user interface is even more important than in an auction with professional bidders. Finally, the auction needs to be tractable, given the fact that it has to be completed in a single day.

A user interface has been developed by an IT company for the bidders to express their preferences. This system shows them the allotment of the solid, including the positioning of the shafts, doors, and windows, and allows them to select lots to form a solid space. If the selected lots do not form a valid solid space, the system immediately provides feedback. Furthermore, a user can provide the user interface with a number of desired characteristics (e.g. surface area, orientation, etc.) and receive a list of solid spaces that satisfy them. This can be done weeks before the start of the auction; interested bidders also have the opportunity to visit the solid beforehand to actually see the space(s) that they intend to bid on. At any time, bidders can bid on at most 8 distinct solid spaces. We found that allowing more different solid space for a bidder to bid on, not only increases the computation time to solve the auction, but also makes it difficult for the bidders to keep track of their bids during the auction. Moreover, since tenants need to provide the interior for their solid space themselves, Stadgenoot is convinced a bidder will not have many different solid spaces in his mindset anyway.

We use a first price system: a bidder pays exactly the amount he bids for the solid space he wins. This is not an obvious choice, since e.g. second price auctions (i.e. the highest bidder pays the price of the second highest bid) have been shown to give each bidder the incentive to bid his true valuation [13]. Nevertheless, the simplicity of the first price concept, taking into account the inexperience of the bidders in the solid auction, and the computational complexity of the Vickrey-Clarke-Groves (VCG) auction (i.e. the combinatorial variant of the second price auction) lead us to adopt a first price rule.

Ideally, an auction ends when each winning bidder is happy with what he wins, and each losing bidder realizes that he is unwilling or unable to pay the amount needed to obtain what he wants. In a first price auction, it suffices that no bidder bids more than he is willing to pay to realize the first part; the second part, however, cannot be guaranteed if the number of rounds is determined before the start of the auction. Nevertheless, for practical reasons, announcing the number of rounds beforehand seems inevitable in our case. Moreover, since Stadgenoot wants the auction to take at most one day, 8 rounds seems to be a maximum (taking into account one hour per round to collect bids and compute the (provisional) winning allocation). We opted for 5 rounds; this choice is backed by our experiments which show that beyond 5 rounds, the total revenue does not rise drastically, provided that bidders have the incentive to start bidding early in the auction.

Because of the limited number of rounds, and Stadgenoot's aspiration for auction stability, it is important that bidders do not wait until the final round to reveal their preferences. We use two rules that encourage the bidders to bid seriously early in the auction. Firstly, a bidder is only allowed to bid on new solid spaces in the first two rounds of the auction; after that, he can only raise his existing bids. Since bidders rent empty shells, and given the fact that it takes a considerable effort for the bidders to plan and budget how they will provide the interior, Stadgenoot is convinced that most bidders have a strict and limited list of solid spaces in which they are interested. Since this list, for most bidders, does not depend on the course of the auction, Stadgenoot believes that it should force the bidders to disclose this information early in the auction. This forces the bidder to disclose the solid spaces in which he is interested already in the first half of the auction, but still gives him some opportunity to focus on solid spaces that turn out to be unpopular after the first round. Secondly, non-winning bidders should raise their bids by a minimum of 0.5 euro per square meter. This rule does not apply to bidders that won a solid space in the previous round: in the current round, they are allowed to wait and see what happens. As stated in section 2, there is a reserve price of 6 euro per square meter. Further, no bidder is allowed to lower or cancel his bid.

In an auction over multiple rounds, it is crucial that bidders receive feedback after each round. With the right feedback, bidders know whether it makes sense to raise their bid, and by how much, whether they should focus on other solid spaces, or whether the auction is beyond their budget. On the other hand, feedback may also encourage collusion or violate the bidder's privacy. Stadgenoot chose to report the winning allocation after each round, together with for each lot the number of bids in which it is involved. The former shows which lots are unallocated, how the solid is divided into solid spaces, and the prices; the latter makes it easy to find out for which lots the competition is likely to be stiff. Although there are ways to provide non-winning bidders with advice about a successful bid raise, Stadgenoot will not offer this kind of feedback. It was felt that this information could be interpreted by the bidder as holding some guarantee of becoming a winner in the next round, which could make Stadgenoot vulnerable to legal claims afterwards. Furthermore, tractability becomes an issue when for each bid a problem at least as hard as the winner determination problem itself needs to be solved.

4 Winner Determination Problem

The problem of deciding which bidders should get what items in order to maximize the auctioneer's revenue is called the winner determination problem. In general, this problem is NP-hard (van Hoesel and Müller [12]) and does not allow good approximation results (Sandholm [10]). However, for some special cases that involve an ordering of the items, the winner determination problem becomes easier.

Rothkopf et al. [9] found that the winner determination problem can be solved in polynomial time if the subsets on which a bidder can bid is limited to hierarchical subsets, i.e. if every two subsets are disjoint or one is a subset of the other. The same goes for settings where a linear order exists among the items and bidders can only bid on subsets of consecutive items, even when the first item in the ordering is considered the successor of the last (i.e. a circular order). This setting is somewhat similar to ours, where the lots are not on a line, but nevertheless arranged in a two-dimensional pattern, and where bidders can only bid on sets of adjacent lots. However, the results of Rothkopf et al. [9] assume a combinatorial auction where a bidder is allowed to win multiple subsets, which is clearly not the case in the solid auction. Nisan [8] elaborates on some of the results of Rothkopf et al. [9] by stating explicitly that the LP-relaxation of a set packing formulation results in an integral solution for the special cases with hierarchical subsets and linearly ordered items. Tennenholtz [11] presents a combinatorial network auction, which he proves is computationally tractable. In this auction, the items are assumed to be arranged in a tree, where every node corresponds to an item. The idea is that bids can be submitted only on subsets of items that form a path in the network. If the items are structured in a directed acyclic graph and bids are allowed on any directed subtree, Sandholm shows that the winner determination problem already becomes NP-hard again [10]. Finally, Day and Raghavan [4] describe the so-called *matrix bid auction*, where each bidder must submit a strict ordering (or ranking) of the items in which he is interested. Goossens et al. [6] show that the winner determination problem for this auction can be solved in polynomial time, given a fixed number of bidders, and provided that all bidders have the same ordering of the items (see also [5]).

We now develop a mathematical formulation for our winner determination problem, based on a set partitioning formulation. We use the following notation. The solid, with a total surface area of A square meters, is divided into a set of lots denoted by K, and has a number of utility shafts $s \in S$. Each shaft s has a ventilation capacity of V_s m^3/h, and offers G_s high capacity gas connections, and E_s high current electricity connections. Per floor $f \in F$ above the ground level, there are two wings $w \in W$, which have a rescue capacity for $O_{f,w}$ persons. At most R restaurants are allowed. We use B_r, B_c, and B_s to represent the set of residential, commercial and social bidders, and they should be awarded at least a fraction f^r, f^c, and f^s of the surface area of the solid respectively. $B = B_r \cup B_c \cup B_s$ denotes the set of all bidders. Each bid $t \in T$ belongs to one bidder $b(t)$, and is characterized by the following parameters. We use $L(t)$ to represent the set of lots included in bid t, a_t for the surface area of this solid space, and p_t for the price that the bidder is willing to pay for it as monthly rent. The solid space is situated on wing $w(t)$ and floor $f(t)$, and the utility shaft that is to be used is given by $s(t)$. If the bidder does not explicitly mention which shaft he wishes to use, we duplicate his bid for each shaft that is contained in his solid space. We use o_t to denote the number of persons for which rescue capacity is needed for bid t, and v_t for the required ventilation capacity. We define T^{R} as a subset of the set of bids T containing those bids where the solid space is to serve as a restaurant. We also have subsets of T for those bids requiring a high-capacity gas (T_s^{G}), and/or electricity (T_s^{E}) connection on shaft $s \in S$. We also added two dummy bidders $d_s, d_r \notin B$, where the former bidder bids on every (valid) solid space small enough for social bidders, and the latter bids on all other (valid) solid spaces. Both bidders always bid a price of zero, and require some minimal ventilation and rescue capacity. We use the decision variable x_t which is 1 if bid t is allocated, and 0 otherwise.

maximize

$$\sum_{t \in T} p_t x_t \tag{1}$$

subject to

$$\sum_{t \in T: k \in L(t)} x_t = 1 \qquad \forall k \in K \tag{2}$$

$$\sum_{t \in T: b(t) = b} x_t \leqslant 1 \qquad \forall b \in B \tag{3}$$

$$\sum_{t \in T: b(t) \in B_r} a_t x_t \geqslant f^r A \tag{4}$$

$$\sum_{t \in T: b(t) \in B_c} a_t x_t \geqslant f^c A \tag{5}$$

$$\sum_{t \in T: b(t) \in B_s \cup \{d_s\}} a_t x_t \geqslant f^s A \tag{6}$$

$$\sum_{t \in T^{\mathrm{R}}} x_t \leqslant R \tag{7}$$

$$\sum_{t \in T: w(t) = w, f(t) = f} o_t x_t \leqslant O_{f,w} \qquad \forall w \in W, f \in F \tag{8}$$

$$\sum_{t \in T: s(t) = s} v_t x_t \leqslant V_s \qquad \forall s \in S \tag{9}$$

$$\sum_{t \in T_s^{\mathrm{G}}} x_t \leqslant G_s \qquad \forall s \in S \tag{10}$$

$$\sum_{t \in T_s^{\mathrm{E}}} x_t \leqslant E_s \qquad \forall s \in S \tag{11}$$

$$x_t \in \{0, 1\} \qquad \forall t \in T \tag{12}$$

The objective function (1) states that the total rent should be maximized. The first set of constraints (2) enforces that each lot needs to be allocated. Indeed, this is necessary to ensure that unoccupied lots (i.e. lots allocated to a dummy bidder) form a valid solid space. The second set of constraints (3) ensures that each bidder wins at most one solid space (except for the dummy bidders). Constraints (4) - (6) make sure that each type of bidder acquires at least a given percentage of the total surface area, the next constraint prevents that more than R restaurants get a place in the solid. Constraints (8) guarantees that the rescue requirement is available for each wing, and constraints (9) enforce that the ventilation capacity is respected for each shaft. Constraints (10) - (11) enforce that per shaft, no more than the available high capacity gas and electricity connections are used. The final set of constraints makes sure that bids are fully accepted or not at all.

Since we have dummy bids on all valid solid spaces, constraints (3) can always be satisfied by allocating the entire building to dummy bidders. Similarly, by including the social dummy bidder d_s in constraint (6), this constraint will not cause infeasibility. However, due to constraints (4) and (5), there may not be a feasible solution. If this happens in the first round, the round is recomputed with f^r lowered to 15%, and if necessary, again with both f^r and f^c lowered to 0%. In the latter case, the model will produce a feasible solution (e.g. leaving the whole building empty), but it will be announced to the bidders that in the current setting, the conditions for a valid auction result are not satisfied. Bidders can use the second round to bid on other solid spaces, such that after the second round, constraints (4) and (5) are satisfied. Otherwise, the model returns infeasibility, and the auction is canceled. In the next section, we implement this formulation and we discuss a number of computational experiments.

5 Computational Experiments

We implemented the formulation described in section 4 and solved it using IBM ILOG Cplex, version 12. In order to evaluate the performance of our algorithm, we carried out a number of computational experiments on randomly generated instances for the auction of Solid 11. Stadgenoot performed several studies with the intention of gaining information about interested tenants, e.g. in which solid spaces they are interested, and what special needs they have (ventilation, electricity, gas, etc.). They expect 55% social bidders, 30% residential bidders, and 15% commercial bidders. In order to have an idea of the amounts they would be willing to bid, Stadgenoot looked at prices for similar apartment spaces nearby. We used this information to generate realistic instances with 50, 150, 500, 1000, and 2000 bidders, where each bidder expressed between 4 and 8 bids. The main goal of these experiments is to evaluate whether the computation times are reasonable. Therefore, only one single round was considered. All instances were solved on a Windows XP based system, with 2 Intel Core 2.8GHz processors. The results are summarized in Table 1; each line gives values that are averaged over 10 similar instances.

The first column shows the number of bidders; the second gives the total number of bids (without the dummy bids). Stadgenoot was unsure about the number of participants, however, having over 1000 bidders was not considered unlikely. The total rent (in euros) is given in the third column. The increase in total rent gradually diminishes, suggesting that as soon as 500 bidders are participating,

Table 1. Overview of computational experiments

Bidders	Bids	Total rent	Unallocated space	Avg. comp. time	Max comp. time
50	299	69,593	21.2%	0.5	1.6
150	913	87,966	5.0%	0.9	1.8
500	2,995	101,824	0.0%	4.0	8.8
1000	6,007	105,838	0.0%	26.1	172.7
2000	11,981	107,018	0.0%	100.3	451.9

the auction will not yield much excess value from extra bidders. Furthermore, the fourth column shows that as soon as 500 bidders can be reached, the solid is fully allocated. With only 50 bidders, as much as one fifth of the building remains empty. Nevertheless, we should take into account that only one round was considered in these experiments. We may expect the final percentage of unallocated space to be lower (as bidders will no doubt focus on unallocated solid spaces in the next round) and the total rent to be higher (due to competition in later rounds). The next column gives the average computation time (in seconds), followed by the maximal computation time in the final column. Clearly, these instances can be solved very efficiently, although there are considerable differences in computation time between instances with the same number of bidders. From these experiments, we conclude that we may expect our model to compute an optimal allocation within the range of 15 minutes (900 seconds) that was postulated in section 3.

6 Conclusions

We described a real-life combinatorial auction for allocating real estate. Due to the concept of solids, a variety of preferences and usages has to be accommodated. Also, different restrictions need to be taken into account in the winner determination problem. We developed a formulation that allows to allocate solid spaces to bidders efficiently. Based on computer simulations and experiments with human bidders, a multi-round auction format with limited feedback was chosen, where bidders are encouraged to start bidding early in the auction.

References

1. Solids: an introduction, `http://www.solids.nl/english`
2. Bijdendijk, F.: Met andere ogen. Het Oosten Woningcorporatie, Amsterdam (2006)
3. Cramton, P., Steinberg, R., Shoham, Y.: Combinatorial auctions. MIT Press (2005)
4. Day, R.W., Raghavan, S.: Matrix bidding in combinatorial auctions. Operations Research 57(4), 916–933 (2009)
5. Goossens, D.R., Müller, R., Spieksma, F.C.R.: Algorithms for recognizing economic properties in matrix bid combinatorial auctions. INFORMS Journal on Computing 22(3), 339–352 (2010)

6. Goossens, D.R., Spieksma, F.C.R.: Exact algorithms for the matrix bid auction. Computers and Operations Research 36, 1090–1109 (2009)
7. Goossens, D.R., Spieksma, F.C.R.: Design Study Solids-veiling (2010) (Confidential, in Dutch)
8. Nisan, N.: Bidding and allocation in combinatorial auctions. In: EC 2000: Proceedings of the 2nd ACM Conference on Electronic Commerce, Minneapolis, USA, pp. 1–12. ACM Press (2000)
9. Rothkopf, M.H., Pekeç, A., Harstad, R.M.: Computationally manageable combinational auctions. Management Science 44(8), 1131–1147 (1998)
10. Sandholm, T.: Algorithm for optimal winner determination in combinatorial auctions. Artificial Intelligence 135(1-2), 1–54 (2002)
11. Tennenholtz, M.: Tractable combinatorial auctions and b-matching. Artificial Intelligence 140(1-2), 231–243 (2002)
12. Van Hoesel, S., Müller, R.: Optimization in electronic markets: examples in combinatorial auctions. Netnomics 3(1), 23–33 (2001)
13. Vickrey, W.: Counterspeculation, auctions, and competitive sealed tenders. Journal of Finance 16, 8–37 (1961)

A Computational Monetary Market for Plug-In Electric Vehicle Charging

Matteo Vasirani and Sascha Ossowski

Centre for Intelligent Information Technology,
University Rey Juan Carlos,
Madrid, Spain
{matteo.vasirani,sascha.ossowski}@urjc.es

Abstract. The penetration of plug-in electric vehicles (PEV) into near-future traffic and power infrastructure is expected to be large enough to have a serious impact on the grid. If each PEV arrives at home and charges immediately, the distribution network can incur in serious problems. Therefore, the charging process of the PEVs has to be coordinated, on the basis of the grid capabilities, generation and pricing. In this paper, we put forward a computational monetary market intended as an automatic scheduler for the charging problem. The market is designed so as i) to satisfy the constraints of the distribution network, ii) to guarantee a reasonable level of fairness and allocation efficiency, while at the same time iii) to give the possibility to each PEV to transiently increase its share of the charging capacity of the local distribution network when needed.

Keywords: Monetary markets, lottery scheduling, smart grid, plug-in electric vehicles, exchange rates.

1 Introduction

The rate at which plug-in electric vehicles (PEV) enter the vehicle stock depends on many factors, including battery cost and reliability, the price of gasoline, and government incentive programs. However, the penetration of these vehicles is expected to be large enough to have a serious impact on the grid. For instance, in [5] it has been estimated that for some scenarios where two PEVs are charging at the same time on the same distribution network, the addition of a hair dryer on the same network will seriously stress it. If each PEV arrives at home and charges immediately, the distribution network can incur in serious problems, since historically it has not been designed for that kind of electricity intensive loads. Therefore, the charging process of PEVs has to be coordinated, on the basis of the grid capabilities, generation and pricing.

One way to address this problem is using time-of-use (TOU) pricing plans to dissuade PEV owners from charging at peak times, when the distribution network the PEV is plugged to is already close to the maximum capacity. However, such an approach in inherently static and rely on the response of the human

P. Coles et al. (Eds.): AMMA 2011, LNICST 80, pp. 88–99, 2012.

owners, who should adapt their behaviour and manually change the time at which their PEVs have to be charged.

A more advanced way to address the charging problem is automatic scheduling. With automatic scheduling we envision a system where PEVs are equipped with intelligent agents that are able to negotiate with some infrastructure component that controls the local distribution network (e.g., a transformer) in order to charge their batteries, while satisfying the constraints of the distribution network.

The paper is structured as follows. In Section 2, we detail a list of properties that our resource management need to satisfy. In Section 3 we discuss some related works and their limitations. Section 4 formalises the PEV charging problem. In Section 5 we propose our resource management mechanism for PEV charging, which is experimentally evaluated in Section 6. Finally, we conclude in Section 7.

2 Desiderata

In designing a mechanism for the automatic scheduling of PEV charging, we aim to satisfy the following list of properties:

Starvation-free. The scheduling mechanism must guarantee that all PEVs have a non-zero probability of charging, at least partially, their batteries.

Fairness and competitiveness. Since different PEVs have different needs, related to the quantity of electricity they demand or when the electricity is needed, the scheduling mechanism should give the PEVs the possibility of increasing their share of the distribution network capacity when needed. However, it is desirable that no agent can unlimitedly increase this share and that a fair share of the distribution network capacity is statistically guaranteed to every PEV.

Environments. Power systems are nowadays highly regulated, and in general consumers pay electricity according to fixed per unit price plans. The proposed scheduling mechanism must work in regulated environments, where electricity is paid at a fixed price, as well as in deregulated environments, where dynamic pricing [3] can be used so as to influence the consumers decision making.

3 Related Work

The problems and challenges posed by the electrification of urban transportation are quite recent. Therefore, the application of agent-based techniques as the means of tackling these challenges is still in its infancy.

In [6], an agent-based solution for the automatic scheduling of PEV charging is presented. This work allows PEV owners to specify the time at which the PEV will be available for charging, the quantity of energy that must be stored in the battery, as well as the time at which the PEV is needed to be charged.

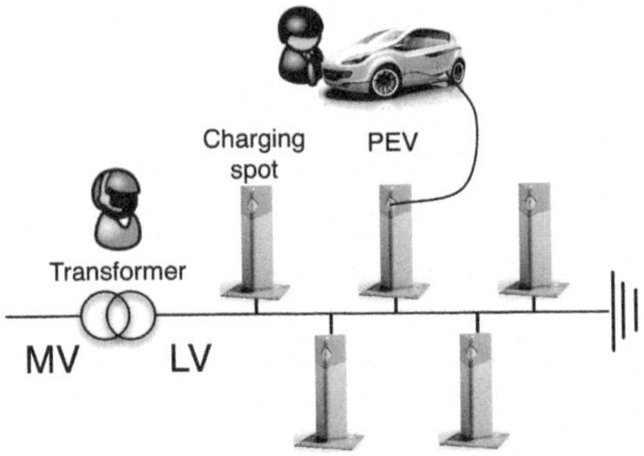

Fig. 1. Distribution network with charging spots

The PEV agent communicates this information to the agent that resides on the local transformer, which performs the automatic scheduling while satisfying the constraints of the distribution network. In general, such an approach is unable to dissuade owner to misreport information. For instance, an owner may indicate an earlier departure time or further travel distances in order to receive preferential charging. However, if all the agents are truthful, it implicitly guarantees a fair charging, since the scheduling is performed by the transformer agents that do not have preferences with respect to the PEVs.

A way to address the above shortcomings is allowing dynamic pricing [3] and defining incentive compatible market clearing mechanisms, such as in [4]. These mechanisms aim at providing PEV agent the monetary incentives to truthfully report their demand vector and the vehicle availability, regardless of the others' reports. However, this approach has two main shortcomings. The mechanism occasionally requires units of electricity to be "burned", either at the time of allocation or on departure of the agent. The former way of "burning" is more realistic but it makes the computation of the allocation hard. The latter is easier to compute, but requires the battery to be discharged on departure, which may not always be feasible in practice, since discharging is not instantaneous and the charging spot must physically impede the PEV owner to unplug the PEV. Furthermore, such an approach is strictly dependent on the possibility of applying dynamic pricing, while nowadays power systems are highly regulated and consumers typically pay electricity according to fixed per unit price plans.

4 PEV Charging

We assume a typical configuration of a local distribution network (see Fig. 1). A transformer device converts the voltage from medium to low. The low voltage feeds several charging spots (from a dozen to one hundred spots), where PEVs connect to charge their batteries. We assume that both the local distribution transformer and the PEVs are equipped with intelligent software agents.

4.1 Transformer Agent

The transformer agent of the local distribution network is in charge of activating the spots at which PEVs are connected to enable charging. As part of the infrastructure, we assume that the system designer is able to program the behaviour of this agent, as well as the scheduling mechanism that regulates which charging spot will be activated. We consider a model with discrete time steps $t \in \mathcal{T}$ with fixed duration Δ. For sake of simplicity, we assume that at each time step only one charging spot can be activated.

4.2 PEV Agent

Let $\mathcal{V} = \{1, 2, \ldots, n\}$ be a set of PEVs. For every PEV, let $c \in \mathbb{R}^+$ be the quantity of electricity demanded by the PEV, which for simplicity is equal to the capacity of the battery, $a \in \mathcal{T}$ the time slot when the PEV is plugged to the charging spot, and $d \in \mathcal{T}$ the departure time slot (i.e., the time slot at which the PEV need to have charged the demanded quantity in the battery). Given the charge rate γ, it is possible to compute the number of time steps $k = c/(\gamma \cdot \Delta)$ at which the PEV needs to be charging. The objective of each PEV is therefore having its charging spot activated for k time slots after its arrival and before its departure.

5 Resource Management for PEV Charging

The resource management mechanism we propose is inspired by *lottery scheduling* [7], a randomised resource allocation mechanism that has been developed for resource allocation in operating systems. In lottery scheduling resource rights are represented by lottery tickets, and an allocation is determined by holding a lottery. The agent with the winning ticket is granted the resource. This effectively allocates resources to competing agent in proportion to the number of tickets that they hold. Lottery tickets are denominated in different *currencies* (one for each PEV), and each currency is backed by a *base commodity*. Ultimately, lottery tickets are equivalent to monetary notes, which are issued in different denominations and are backed by a base commodity (e.g., gold).

5.1 Registration

When a PEV is plugged in the charging spot for the first time, the corresponding PEV agent registers itself with the local transformer agent. The PEV agent sends a `register` message, providing its identification number, assumed to be unique. The transformer agent confirms the registration, providing the initial monetary base m of the PEV agent which is determined as follows. Let G be an arbitrary constant integer value. This value represents the overall quantity of base commodity that backs the different currencies[1]. Let q be the number

[1] In monetary terms, G is equivalent to the quantity of gold in circulation.

of charging spots in the local distribution network. The base commodity G is *equally split* among the PEVs, so as that each PEV owns a share $g = G/q$ of the base commodity. This share funds the monetary base m of the PEV agent, according to the exchange rate r. The exchange rate determines the quantity of base commodity a monetary note is worth, according to the equation $m = r \cdot g$

In the registration phase, the initial exchange rate r is set to 1, and therefore the monetary base m is equal to g. We remark that while G and g are constant, the monetary base m and the exchange rate r change with time[2]. In particular, the monetary base can be expanded by the PEV agent, which issues the monetary notes, while the exchange rate is set by the transformer agent. The transformer agent maintains a table with the actual values of the base commodity g, the monetary base m with its currency[3], and the exchange rate r.

5.2 Lottery

As said before, we assume a model with discrete time steps $t \in \mathcal{T}$ with fixed duration Δ. This means that at each time step one of the charging spots with a connected PEV must be activated to enable charging for the whole duration of the time step. Thus, at the beginning of each time step, the transformer agent sends a proposal, contained in a `call_for_lottery` message, to all the registered PEVs. A PEV that wants to participate in the lottery replies with an `accept` message and reports the monetary base m, which determines the probability that its charging spot will be activated. Since the PEV agent control the monetary base, it is possible that the new monetary base differs from the actual monetary base that is stored by the transformer agent. In fact, a PEV may try to increase its relative probability by *inflating* its monetary base.

The transformer agent then updates the exchange rate of all the PEVs. Let $M^t = \sum_{i \in \mathcal{V}} m_i^t$ be the sum of the PEV agents' monetary base submitted at time t, G the total base commodity, $\eta \in (0, 1]$ a tunable parameter, and $H(\cdot)$ the Heaviside step function, whose value is zero for negative argument and one for positive argument. The PEV exchange rates is updated according to Eq. 1, where $\alpha(m^t, m^{t-1}) : \mathbb{R}^+ \times \mathbb{R}^+ \to (0, 1]$ is an update step function, which depends on the new monetary base m^t, submitted by the PEV agent, and the old monetary base m^{t-1}, stored by the transformer agent. The rationale of this exchange rate update scheme will be detailed in section 5.3.

$$r \longleftarrow r + \alpha(m^t, m^{t-1}) \cdot \left(\frac{m^t}{g} - r\right)^{1+\eta H\left(\frac{m^t}{g} - r\right)\left(\frac{M^t}{G} - 1\right)} \quad (1)$$

After the update of the exchange rate, also the monetary base of each PEV that is stored by the transformer agent is updated to the new value m^t. Then the lottery takes place and the transformer agent draws the winner ticket according

---

[2] When necessary, we will use the notation m^t and r^t respectively to express the dependence of time of these quantities.

[3] For simplicity, the currency is set to the ID of the PEV.

to Eq. 2. The ratio between the monetary base and the exchange rate gives the probability of a charging spot to be activated.

$$p(i) = \frac{\frac{m_i}{r_i}}{\sum_{j \in \mathcal{V}} \frac{m_j}{r_j}} \tag{2}$$

The transformer agent probabilistically selects the winner of the lottery and notifies the participants about the outcome. After that, the transformer agent physically activates the charging spot to which the winner PEV is connected. The PEV then starts charging the battery for the entire duration of the time slot, and it is charged for the electricity that it draws from the grid according to a fixed per unit price plan.

5.3 Exchange Rate Update

As said before, the monetary base m of a PEV is related to the base commodity g through the exchange rate r. For example, let's suppose that the exchange rate is 1. In this case, the value of one monetary note equals the value of one unit of the base commodity. If a PEV agent expands its monetary base by doubling the number of notes, the value of the monetary note should halve, since the base commodity g that funds the monetary base is constant. However, if the inflationary adjustment of the exchange rate were instantaneous, an expansion of the monetary base performed by a PEV agent would not have any positive effect on the probability of having its charging spot activated: if m is doubled and r is halved, the probability of winning the lottery remains constant (see Eq. 2).

For this reason, we perform a *delayed update* of the exchange rate r, which enables the PEV agents to *transiently increase* the probability of having its charging spot selected. In this way, a PEV that needs urgent charging may try to improve the probability of actually being allowed to charge. Analysing Eq. 1, we can distinguish three cases:

1. When the actual exchange rate r is lower then m/g, it means that the exchange rate is being adjusted towards the "true" exchange rate as a reaction to the expansion of the monetary base performed by the PEV agent. In this case, the update step function α plays a central role. A value of α close to 0 considerably delays the inflationary adjustment. On the other hand, a value of α close to 1 implies an almost instantaneous inflationary adjustment, which in practice impedes the inflationary agent to actually increase its probability of being able to charge. In this work, we implement a relative update step function. With this function, the update step increases with the expansion of the monetary base (see Eq. 3). The minimum update step is set to a constant value $\beta \in (0, 1]$, while the degree of increase is determined by another constant parameter $\rho \in \mathbb{R}^+$

$$\alpha(m^t, m^{t-1}) = \begin{cases} \beta & if\ 0 < \dfrac{m^t}{m^{t-1}} \leq 1 \\ 1 - (1-\beta)e^{\rho - \rho \frac{m^t}{m^{t-1}}} & if\ \dfrac{m^t}{m^{t-1}} > 1 \end{cases} \quad (3)$$

Additionally, when $r < m/g$ the Heaviside function $H(\cdot)$ returns 1, which implies that the adjustment that is done towards the "true" exchange rate is raised to a power exponent that depends on the ratio between the overall monetary base of all the agents (M^t) and the total base commodity (G). This exponent penalises excessive inflation, which in turn makes the monetary expansion not effective[4]. The constant parameter η implicitly determines from which extent the overall expansion of the monetary base starts to be detrimental for the inflationary agents, actually reducing, rather than increasing, their probability of winning the lottery.

2. When the actual exchange rate r is greater then the "true" exchange rate m/g, it means that the exchange rate is being reduced towards the "true" exchange rate, as a reaction to an eventual deflation of the monetary base performed by the PEV agent. In this case, $\alpha(m^t, m^{t-1})$ is constant and equal to β, and the Heaviside function returns 0, so as the adjustment is raised to the power of 1.
3. When the actual exchange rate r is equal to the "true" exchange rate m/g no adjustment takes place.

6 Experimental Evaluation

In this section, we empirically evaluate our proposed mechanism. The objective of the experiments is assessing *allocative efficiency* of the resource management mechanism, as well as the *social welfare* of the PEV agents.

The allocative efficiency (ϕ) refers to how close to satisfy the aggregated demand is the resource management mechanism. The demanded electricity is represented by the overall capacity of the PEV batteries. Given that only one PEV can charge in a single time slot, satisfying the aggregated demand is not always possible, since it depends on the number of PEVs. Let $s^{\max}$ be the maximum feasible supply, which can be computed by a centralised scheduler with full knowledge of the aggregated demand and the arrival and departure time slots[5]. Finally, let s be the supply provided by the resource management mechanism, given by the summatory over the state of charge of the PEV batteries at the end of the time window. The allocative efficiency is therefore $\phi = s/s^{\max}$.

[4] In fact, expanding the monetary base is effective *only if* a certain subset of the PEV agents acts in that way.

[5] Due to lack of space, we omit the detail of the optimal computation of $s^{\max}$.

There are manifold measures of social welfare, given some utility function u_i, to assess the overall quality of an allocation [2]. Let b_i be the battery state of charge at the end of the time window, and c_i the demanded electricity, i.e., the capacity of the battery. We define the PEV utility as $u_i = b_i/c_i$. The social welfare metrics we use are the utilitarian social welfare, σ_u, and the Nash product, σ_N (Eq. 4). The utilitarian social welfare is simply the average of the utility gained by each PEV. The Nash product is the product of the utilities of each agent. This notion of social welfare favours both increases in overall utility and inequality-reducing redistributions.

$$\sigma_u = \frac{1}{|\mathcal{V}|} \cdot \sum_{i \in \mathcal{V}} u_i \qquad \sigma_N = \prod_{i \in \mathcal{V}} u_i \tag{4}$$

6.1 Experimental Setup

In the experimental setup we assume that charging occurs within a fixed time window of 12 hours (e.g., from 8PM to 8AM of the next day), divided in time slots of $\Delta = 1$ min. Therefore, the set of time slots is $\mathcal{T} = \{1, 2, \ldots, 720\}$. The local distribution network is equipped with fast charging spots that provide a power of 40kW (400V@63A), which is a common standardised socket in European three-phase networks. A PEV is connected to each charging spot, with a battery capacity uniformly drawn from $15 - 25$kWh. This implies that each PEV needs $22.5 - 37.5$ min of charging. For each PEV, the arrival time slot a is drawn from a half-Gaussian probability distribution over the interval $[1, 720]$, with mean equal to 1 and standard deviation of 60 time slots. Similarly, the departure time slot d is drawn from a half-Gaussian probability distribution over the interval $[a, 720]$, with mean equal to 720 and standard deviation of 60 time slots. This means that 68.2% of the PEVs arrive in the first hour of the time window (e.g., from 8PM to 9PM) and departure in the last hour (e.g., from 7AM to 8AM).

As said in section 5.2, at the beginning of each time slot a PEV agent must accept or refuse to participate in the lottery. After receiving the `call_for_lottery` message, the PEV agent checks the state of charge of its battery and if the battery is not full it sends an `accept` message, which includes the reported monetary base. The real decision making of the PEV agent relies on the selection of the monetary base included in the `accept` message. In this evaluation, we conceive two different strategies: `Zero-Intelligence` and `Some-Intelligence`. With the `Zero-Intelligence` (ZI) strategy a PEV agent simply submits the monetary base that was set in the registration phase. With the `Some-Intelligence` (SI) strategy the PEV agent inflates its monetary base with a certain probability p_d:

$$p_d = \left(1 - \frac{d - t}{d}\right) \tag{5}$$

where t is the time slot at which the lottery is taking place and d is the departure time slot. The rationale of the probability p_d is that the closer the PEV is to the departure, the higher the probability of inflating the monetary base, trying to increase the likelihood of winning the lottery.

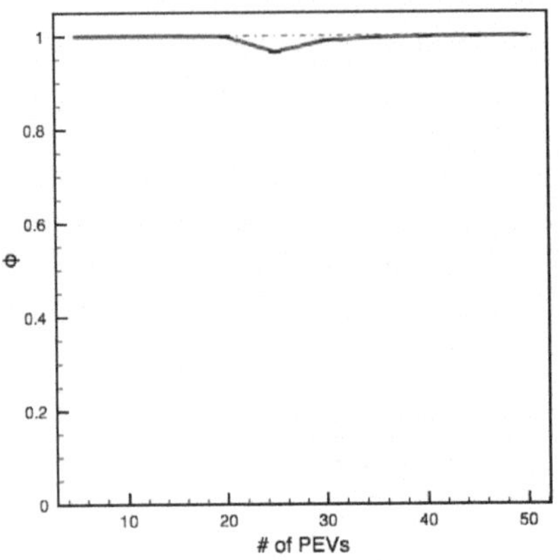

(a) Allocative efficiency

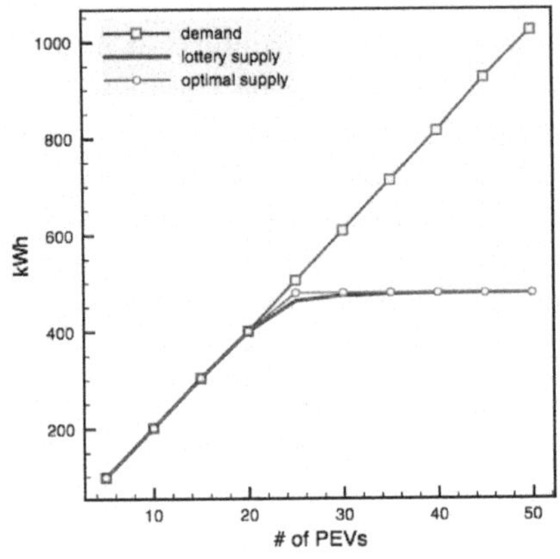

(b) Demand and supply

Fig. 2. Efficiency metrics

We perform experiments for different values of β, ρ and η for the exchange rate update step function, and different compositions of the set of PEV agents, according to their strategies. For each experimental configuration, we run 100 trials and we compute average values of the metrics of interest. All error-bars denote 95% confidence intervals.

6.2 Results

Fig. 2(a) shows the allocative efficiency of the resource management mechanism. These results have been obtained for $\beta = 0.1$ and $\rho = 0.5$, although similar dynamics have been obtained with different combinations of these two values. It is noticeable how the mechanism guarantees a quite high allocative efficiency. Furthermore, as the number of PEVs increases, the allocative efficiency tends to 1, since the highest feasible supply is reached when more than 25 PEVs are connected to the distribution grid (see Fig. 2(b)). To evaluate the allocative efficiency we used a population entirely composed of ZI agents. We remark that the specific strategy followed by the PEV agent does not affect the value of the aggregated demand that is satisfied, but rather how the aggregated demand is individually satisfied among the agents.

Additional interesting insights can be derived from the analysis of the social welfare metrics. In these experiments, we are interested in determining how the utility of an allocation is distributed among the agents, depending on the strategies followed by PEV agents. Therefore, we run experiments with different percentages of SI agents of the total population of PEVs, while the rest of the agents follow the ZI strategy.

Fig. 3(a) to 3(d) plots the utilitarian social welfare for different compositions of the set of PEV agents. These results have been obtained for $\beta = 0.1$, $\rho = 0.5$

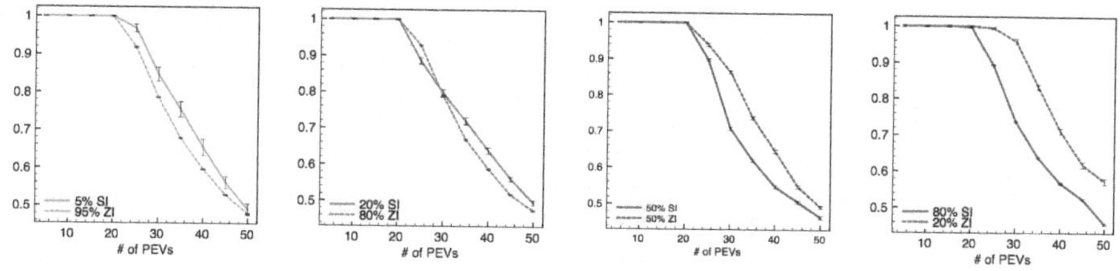

Fig. 3. Utilitarian social welfare

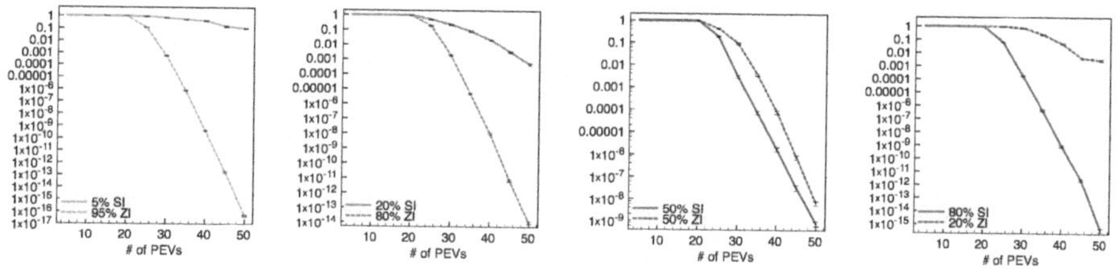

Fig. 4. Nash product

and $\eta = 10^{-27}$. It is possible to notice how the SI strategy is not beneficial when 50% of the agents use that strategy. If too many agents opt to play the SI strategy, the excessive inflation of the overall monetary base makes playing that strategy detrimental, since inflating the monetary base actually decreases the probability of having the PEV's spot selected for charging.

Fig. 4(a) to 4(d) plots the Nash product for the same experimental configuration. Again, the SI strategy does not perform well when 50% of the agents (or more) play that strategy. Therefore, given the parametrisation used in these experiments, there exists a particular splitting of the agent into two groups, ZI agents and SI agents, which represents an equilibrium condition, under which the utility gained by the SI agents must equal the utility gained by the ZI agents.

6.3 Learning to Select the Best Strategy

As highlighted by the social welfare analysis of the previous section, it is clear that PEV agents are faced with the problem of selecting a suitable strategy. Let's suppose the restricted case in which each agent must pick on of the two strategy, ZI or SI, when the PEV is plugged to the charging spot. If we consider a one-shot encounter (i.e., a single charging period, from 8PM to 8AM of the next day), it is hard for a PEV agent to speculate about the best strategy. In fact, playing the SI strategy may not be beneficial if many other PEV agents do the same, as a sort of minority game [1].

Therefore, we setup a learning scheme that is used by the agents to select, at the beginning of each charging time window, the strategy to follow. Let $\overline{u}(y)$ be the current estimate of the agent utility that is obtained if strategy $y \in \mathcal{Y}$

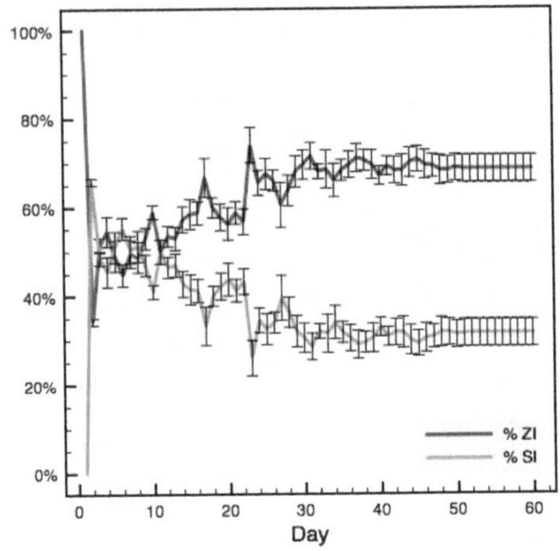

Fig. 5. Learning to pick the best strategy

is followed, where $\mathcal{Y}$ is the set of possible strategies. When the PEV is plugged to the charging spot, the PEV agent selects the charging strategy according to the Boltzmann probability distribution π (see Eq. 6), where $T \in (0, 1]$ is the "temperature" parameter used to enforce convergence over time.

$$\pi(y) = \frac{e^{\overline{u}(y)/T}}{\sum_{y' \in \mathcal{Y}} e^{\overline{u}(y')/T}} \tag{6}$$

After having selected strategy y, we assume that for the entire duration of the time window the PEV agent will adhere to this strategy. At the end of the time window, the PEV agent computes the utility that it obtained and updates the estimate using the formula:

$$\overline{u}(y) = \overline{u}(y) + \frac{1}{w+1}\left(\frac{b}{c} - \overline{u}(y)\right) \tag{7}$$

where w is the number of times the strategy y has been selected so far, b is the battery state of charge at the end of the time window, and c is the available capacity.

To evaluate the outcome of such a learning scheme, we simulate a scenario with 40 PEVs, whose strategy space is $\mathcal{Y} = \{\text{ZI,SI}\}$. At the beginning, each PEV agent is assumed to play the ZI strategy. We also assume that for each PEV agent the initial estimate is $\overline{u}(y) = 1 \ \forall y \in \mathcal{Y}$. As before, we set $\beta = 0.1$, $\rho = 0.5$ and $\eta = 10^{-27}$. The temperature T is initialised to 1 and reduced geometrically by 0.9 every time window.

The result of the experiments are shown in Fig. 5. Using the simple learning scheme described above, the PEV agents selfishly learn to pick the best strategy.

The result is that they split into two groups, with approximately 70% of them playing the ZI strategy, and the remaining 30% playing the SI strategy. This result is in line with the results of the social welfare analysis. In fact, we noticed than when 50% of agents start to play the SI strategy, the utility that those agents receive is lower than that received by the ZI agents. However, for every agent that switches from ZI to SI, there is a reduction of the utility surplus. For the particular parametrisation of the system that we used in the experiments, the splitting 30%-70% between SI agent and ZI agents represents the equilibrium condition.

7 Conclusions

The forecast penetration of plug-in electric vehicles into our cities is expected to have a serious impact on the grid. In this context, we put forward a computational monetary market for the management of the charging process of PEVs. The experimental evaluation showed how the monetary market worked as an automatic scheduler that satisfied the constraints of the distribution network, guaranteeing fairness and allocation efficiency. Furthermore, since it offers the possibility of forcing a desired equilibrium between inflationary and not inflationary agents, it preserved the feature of giving the PEV agents the possibility of transiently increase their share of the disputed resource.

Acknowledgments. This research was partially supported by the Spanish Ministry of Science and Education through the project “AT” (CONSOLIDER CSD2007-0022, INGENIO 2010) and “OVAMAH” (TIN2009-13839-C03-02).

References

1. Arthur, W.B.: Inductive reasoning and bounded rationality. American Economic Review 84(2), 406–411 (1994)
2. Chevaleyre, Y., Dunne, P.E., Endriss, U., Lang, J., Lematre, M., Maudet, N., Padget, J., Phelps, S., Rodríguez-Aguilar, J.A., Sousa, P.: Issues in multiagent resource allocation. Informatica 30(1), 3–31 (2006)
3. Faruqui, A., Hledik, R., Tsoukalis, J.: The power of dynamic pricing. The Electricity Journal 22(3), 42–56 (2009)
4. Gerding, E., Robu, V., Stein, S., Parkes, D., Rogers, A., Jennings, N.: Online mechanism design for electric vehicle charging. In: Proceedings of 10th International Joint Conference on Autonomous Agents and Multi-Agent Systems (2011)
5. Rahman, S.: Analysis of the impact of phevs on the electric power distribution network. In: Transmission and Distribution Conference and Exposition (2010)
6. Vandael, S., Boucke, N., Holvoet, T., Deconinck, G.: Decentralized demand side management of plug-in hybrid vehicles in a smart grid. In: Proceedings of 1st International Workshop on Agent Technologies for Energy Systems, pp. 67–74 (2010)
7. Waldspurger, C., Weihl, W.H.: Lottery scheduling: flexible proportional-share resource management. In: Proceedings of the 1st USENIX conference on Operating Systems Design and Implementation, pp. 1–11. USENIX Association (1994)

An Online Mechanism for Multi-speed Electric Vehicle Charging

Valentin Robu[1], Sebastian Stein[1], Enrico H. Gerding[1], David C. Parkes[2], Alex Rogers[1], and Nicholas R. Jennings[1]

[1] University of Southampton, SO17 1BJ, Southampton, UK
{vr2,ss2,eg,acr,nrj}@ecs.soton.ac.uk
[2] Harvard University, Cambridge, MA 02138, USA
parkes@eecs.harvard.edu

Abstract. As plug-in electric vehicles become more widespread, their charging needs to be coordinated, in order to ensure that capacity constraints are not exceeded. This is becoming particularly critical as new fast-charging technologies are being developed that place additional burden on local transformers. To address this problem, we propose a novel online mechanism in which agents representing vehicle owners are incentivised to be truthful not only about their marginal valuations for electricity units, but also about their arrival, departure and maximum charging speeds. The work extends the state of the art in several ways. We develop an online, model-free mechanism that handles multi-unit demand per period, thus accommodating vehicles with heterogeneous and flexible charging speeds; we provide competitive worst-case bounds for our mechanism; finally, we simulate the proposed online mechanism using data from a real-world trial of electric vehicles in the UK, showing that using fast charging leads to significant cost savings.

1 Introduction

Recent advances in battery technology, and pressures to reduce the carbon emissions of transport, have stimulated renewed interest in electric vehicles (EVs). New hybrid designs, equipped with both an electric motor and an internal combustion engine (that can be used to drive or charge the battery), address common worries about the limited range of such vehicles, and EVs are expected to represent close to 10% of all vehicle sales by 2020, according to a recent Gartner report [7]. However, this potential growth has generated concerns that if many of these vehicles are plugged in, and charged simultaneously, they risk overloading local electricity distribution networks (Shao et al[10]).

To address this problem, a number of researchers have begun to investigate mechanisms to schedule the charging of EVs, such that the local constraints of the distribution network are not exceeded. For example, Clement, Haesen & Driesen [1] propose a centralised scheduler which makes optimal use of the network capacity when vehicle owners truthfully report their expected future vehicle use to the system. However, such approaches fail to address the fact that owners will likely misreport this information if it is in their interest to do so (for example, reporting that they require their vehicle earlier than is actually the case to receive preferential charging). To this end, in this paper we use *online mechanism design*, in order to engineer payment mechanisms that provide incentives for these owners to report truthfully their value for receiving electricity, their willingness to wait and their maximum charging rate.

P. Coles et al. (Eds.): AMMA 2011, LNICST 80, pp. 100–112, 2012.

Online mechanism design is an important topic in distributed AI and economics. For example, Parkes and Singh [8] propose an online variant of the Vickrey-Clarke-Groves (VCG) mechanism, which uses an MDP-type framework to predict future arrivals. Unlike their work, the mechanism proposed here is model-free (i.e., assumes no knowledge about the future). Model-free online settings have been considered by Porter [9] and Hajiaghayi et al. [5], who study the problem of online scheduling of a single, reusable resource over a finite time period. We extend this work by considering multi-unit domains, with preferences described by a non-increasing vector of values.

A different approach for dynamic problems is considered by Juda and Parkes [6]. They consider a mechanism in which agents are allocated options (a right to buy) for the goods, instead of the goods themselves. The concept of options is promising, but would need modifications to apply to our setting with perishable electricity units. In addition to theoretical results, several applications have been suggested for online mechanisms, including: the allocation of Wi-Fi bandwidth (Friedman & Parkes [3]), scheduling of jobs on a server (Porter [9]) and the reservation of display space in online advertising (Constantin et al. [2]).

In recent work (Gerding et al. [4]), we propose the first online mechanism to deal with the problem of coordinating the charging of a set of plug-in hybrid electric vehicles (PHEVs) under limited supply. This earlier model, however, has several limitations that we address in this work. First, all vehicles participating in the system are assumed to have the same charging speed. In fact, given the large numbers of competing manufacturers entering this space, it is likely that domestic EV chargers with a wide range of maximum charging speeds will become available.[1] High performance chargers may pose additional burdens on the local transmission network and, for a real-world deployment, the allocation model and market design needs to be able to deal with such a challenge. Moreover, the presence of multiple, asymmetric charging speeds may considerably affect the dynamics of such a market, as they enable different allocations of the limited network capacity to become feasible. Finally, no theoretical worst-case bound for the mechanism was presented, whereas such a guarantee is highly desirable, especially as online allocation leaves some items unallocated to guarantee truthfulness. Against this background, this paper makes the following contributions to the state of the art:

- We develop a novel online mechanism that deals with multi-unit demands *per time step*, by extending the mechanism proposed by Gerding et al. [4] to accommodate heterogeneous and flexible charging speeds.
- We provide competitive bounds on allocative efficiency compared to the optimal offline allocation (which assumes prior knowledge of future arrivals).
- We simulate the proposed mechanism using data from a real-world trial of electric vehicles in the UK. We show that the use of fast chargers can lead to significant savings in fuel consumption, and is beneficial both from the perspective of individual vehicle owners, and for the allocative efficiency of the whole market.

2 Multi-speed EV Charging Model

We consider a setting in which multiple units of electricity are periodically sold at fixed time steps, t (e.g., once every hour). A *unit* of electricity is defined as the amount of

[1] `http://www.pod-point.com/pod-point-home` provides an example of a domestic charger that already offers a 2.5 times speed increase over standard chargers.

kWh when charging at the lowest rate during that interval. Importantly, we assume that the charging rate can be flexible, and each vehicle has a maximum charging rate (which depends on the battery and its charger). We assume units to be indivisible, which means that the charging rate is a multiple of the lowest rate (this is not a limitation since units can be defined to be arbitrarily small). Moreover, we denote by $S(t)$ the supply of electricity available for EV charging, i.e. the number of units to be sold at time t. Note that, since our allocation is essentially greedy (i.e. units are allocated just before they are they are charged and there is no pre-allocation), this allows us to distribute electricity coming from uncertain sources such as a shared renewable generator (e.g. a shared neighbourhood wind turbine). For now, we consider a market for electricity for EV charging that is separate from that for household consumption, and so S(t) can be considered the residual supply once household consumption has been removed. Beyond providing a manageable model, a practical reason for this separation of concern is to protect one neighbour from higher electricity prices for running regular jobs (lightning, domestic appliances) in the case that his neighbours have purchased EVs.

Let $I = \{1, 2, \ldots, N\}$ denote the set of agents, each of which operates on behalf of a single EV and its owner. Vehicles come and go and are not always available for charging. Furthermore, they can have different maximum charging rates, and their owners have different valuations for the electricity. Given this, an agent i's *type* is described by the tuple $\theta_i = \langle \mathbf{v}_i, a_i, d_i, r_i \rangle$, where $\mathbf{v}_i$ is the marginal valuation vector, a_i and d_i are the arrival and departure times (the earliest and latest times that the EV is available for charging) and r_i is the *maximum* charging rate (i.e., the maximum number of units agent i can charge at any time t).

Each element $v_{i,k}$ of $\mathbf{v}_i$ represents the agent's willingness to pay for the kth unit of electricity. We assume non-increasing marginal valuations, i.e., $v_{i,k} \geq v_{i,k+1}$. This is a realistic assumption for PHEVs since the first few units of electricity are always more likely to be used [4]. Furthermore, a_i and d_i define when the agent is present in the market, where a_i is the agent's *arrival* or earliest time the vehicle can be charged, and d_i is the point of *departure*, after which the vehicle is required by the owner.

Given this, a *mechanism* asks the agents to report their types and decides on an appropriate allocation and payment. We denote the reported type by $\hat{\theta}_i = \langle \hat{\mathbf{v}}_i, \hat{a}_i, \hat{d}_i, \hat{r}_i \rangle$. In practice, the arrival report $\hat{a}_i$ is the time at which the owner plugs a vehicle into the electricity network. At the time of arrival, an agent is also required to report its marginal valuation vector $\hat{\mathbf{v}}_i$. The departure report $\hat{d}_i$ is *not* required in advance and it simply represents the time when the vehicle is unplugged from its unit. A vehicle owner may decide to change its (reported) departure time, by simply unplugging her vehicle.

Agents (or their owners) can misreport their availability, for example, by unplugging the vehicle early or plugging in the vehicle some time after arrival to try and get a better price. Also, they could report a lower maximum charging rate or higher valuations. Our aim is to develop a mechanism which is *dominant strategy incentive compatible* (DSIC), i.e., agents are best off reporting $\hat{\theta}_i = \theta_i$, no matter what other agents report. Formally, a mechanism is given by the allocation policy $\pi_i^{\langle t \rangle}(\hat{\theta}_I | \mathbf{k}^{\langle t \rangle}), i \in I$, which determines the number of units allocated to agent i at time t, and payment policy $x_i(\hat{\theta}_{-i} | k_i), i \in I$, which calculates the total payment on (reported) departure of an agent. We denote by $\theta_I = \{\theta_i | i \in I\}$ the types of all agents, and $\theta_{-i} = \{\theta_j | j \in I, j \neq i\}$ all agents except i. Payments are independent of an agent's own report.Here, k_i is agent i's *endowment* (number of items allocated) on departure, and $\mathbf{k}^{\langle t \rangle} = \langle k_1^{\langle t \rangle}, \ldots, k_N^{\langle t \rangle} \rangle$ denotes the *endowment* of all agents at time t, where$k_i^{\langle t+1 \rangle} = k_i^{\langle t \rangle} + \pi_i^{\langle t \rangle}(\hat{\theta}_I | \mathbf{k}^{\langle t \rangle})$.

3 The Online Mechanism

Essentially, our mechanism uses a greedy allocation policy, which allocates available units to the agents with the current highest marginal valuations (given their current endowments). While this is not always optimal (since it does not consider the fact that some agents leave earlier than others), it is generally applicable, as it does not require information about future arrivals or departures (i.e., it is "model free").

A key problem, identified in our previous work [4], is that greedy allocation in online domains sometimes overallocates (given the prices it has to pay, the agent would prefer to get fewer units than it would be allocated). To address this problem, we correct the allocation by leaving some units unallocated under well specified conditions, which we refer to as 'burning' units. We consider two types of burning, discussed below: *immediate* burning, where units are simply left unallocated (i.e., none of the agents receive the unit, even if there is a demand for them), and *on-departure*, where units are initially allocated using the greedy approach (i.e., the battery is being charged), but then on departure of the agent, any overallocated units are discharged from the battery. Both of these have advantages/disadvantages. The model with on-departure burning is more efficient (i.e., generally burns fewer units) and is also more computationally efficient in computing the payments. However, it may not be realistic to expect that we can partially discharge a car's battery on the departure, so the model with immediate burning may be more realistic, given the application.

Formally, let the vector $\mathbf{b_i^{\langle t \rangle}} = \langle \hat{\mathbf{v}}_{\mathbf{i},\mathbf{k_i^{\langle t \rangle}}+1}, \ldots, \hat{\mathbf{v}}_{\mathbf{i},\mathbf{k_i^{\langle t \rangle}}+\hat{\mathbf{r}}_\mathbf{i}} \rangle$ denote agent i's reported marginal values for the next $\hat{r}_i$ units, given its endowment $k_i^{\langle t \rangle}$ at time t. It is convenient to think of this as the agent's *bids* for the units available at time t. Furthermore, let $B^{\langle t \rangle}$ denote the multiset of such bids from all agents that are present in the market at time t, i.e., from all $i \in I$ s.t. $\hat{a}_i \leq t \leq \hat{d}_i$. The allocation rule is:

Definition 1 (Greedy Allocation). *At each time step t, allocate the $S(t)$ units of electricity to the highest bids in $B^{\langle t \rangle}$.*

To ensure that the payment is report-independent, the mechanism reruns the market without agent i (from $\hat{a}_i$ onwards, since agent i did not affect the market prior to this time). Let $B_{-i}^{\langle t \rangle}$ denote the multiset of the bids placed by all agents in the market at time t if agent i were removed and the market were rerun from $\hat{a}_i \leq t$ onwards. In case $|B_{-i}^{\langle t \rangle}| < S(t)$, we add a number of zero-valued bids and refer to this enlarged set as $B_{-i\cup 0}^{\langle t \rangle}$, to ensure that $|B_{-i\cup 0}^{\langle t \rangle}| \geq S(t)$. Next, we define set operators $\max_k B$ and $\min_k B$ to return the highest and, respectively, lowest k elements of multiset B (or, if $|B| < k$, to return B). Then, we define the *externality* that agent i would impose on other agents if it won $\min(r_i, S(t))$ out of $S(t)$ units at time t as:

$$E_i^{\langle t \rangle} = \min_{r_i}(\max_{S(t)} B_{-i\cup 0}^{\langle t \rangle})$$

Intuitively, the multiset $E_i^{\langle t \rangle}$ here contains the bids from other agents that would lose out if agent i were to win r_i units at time t. Note that the intuition here is the same as in the

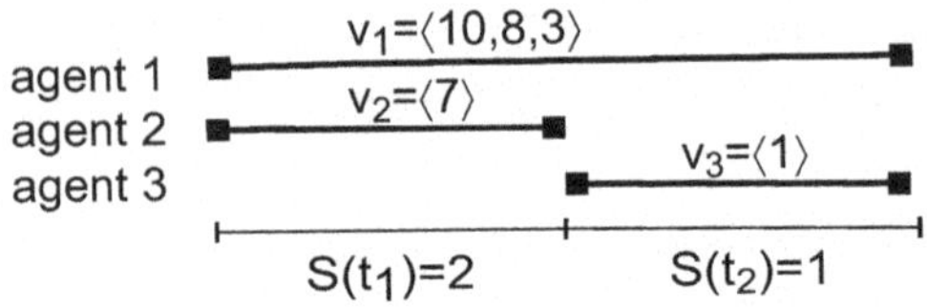

Fig. 1. Example showing arrivals, departures, and valuation vectors of 2 time units and 3 agents

regular Vickrey-Clarke-Groves (VCG) mechanism, as the total payment corresponds to the sum of the externalities. However, to compute the overall payments online, we need to combine these externalities across all time steps in the agent's active period up to current time t. To do this, we define an ordered vector of *prices*, $p_{-i}^{\langle t\rangle}$, as follows:

$$\mathbf{p}_{-i}^{\langle t\rangle} = \text{incr}\left(\bigcup_{t'=a_i}^{t} E_i^{\langle t'\rangle}\right),$$

where incr is an operator that orders elements from a multiset in increasing order, and we use the union symbol to denote the union of multisets (and so the same element can appear multiple times). Now, $p_{-i,k}^{\langle t\rangle}$ is the price that agent i is charged for the k-th unit of electricity. Intuitively, this is the minimum valuation that the agent could report for winning this unit by time t. Note we use $p_{-i,k}$ to denote this price at time $\hat{d}_i$. Given this, the mechanism is as follows.

- **Decision Policy.** The decision consists of two stages.
  Stage 1. At each time point t, *pre-allocate* using Greedy Allocation (see Definition 1).
  Stage 2. We consider two variations of when to decide to burn pre-allocated units:
  - **Immediate Burning.** Leave any unit unallocated whenever the price for this unit is greater than the marginal value, i.e., whenever:

$$\hat{v}_{i,k} < p_{-i,k}^{\langle t\rangle} \text{ for } k_i^{\langle t\rangle} < k \leq k_i^{\langle t\rangle} + \pi_i^{\langle t\rangle}$$

  - **On-Departure Burning.** For each departing agent, discharge any unit $k \leq k_i$ where $\hat{v}_{i,k} < p_{-i,k}$.
- **Payment Policy.** Payment always occurs on reported departure (i.e., when the owner/agent unplugs the vehicle). Given that k_i units are allocated to agent i, the payment collected from i is:

$$x_i(\hat{\theta}_{-i}|k_i) = \sum_{k=1}^{k_i} p_{-i,k} \tag{1}$$

Figure 1 illustrates the mechanism through an example with 2 time steps and 3 agents, showing the agents' arrival, departure and valuations. Suppose furthermore, that supply is $S(t_1) = 2$ and $S(t_2) = 1$. Now, consider 2 distinct cases:

The maximum charging speed of agent 1 is $r_1 = 1$.[2] In this case, at most one marginal value is taken from each agent. At time t_1, marginal valuations $v_{1,1} = 10$ of agent 1, and $v_{2,1} = 7$ of agent 2 are allocated, while at time t_2, marginal value $v_{1,2} = 8$ of agent

[2] Note that the other 2 agents only desire one unit, so their maximum charging speed is irrelevant in this example.

1 is allocated. The prices charged to agent 1 are: $p_{-1} = \langle 0, 1 \rangle$, because without agent 1 in the market, there would be a free, spare unit at time t_1 and the available unit at t_2 would sell to agent 3 for 1. No units get burnt in this case, and the actual allocation is actually equivalent to the optimal offline allocation.

The maximum charging speed of agent 1 is $r_1 = 2$. Then, at time t_1, the greedy mechanism described above allocates the 2 marginal values of agent 1: $v_{1,2} = 10$ and $v_{1,2} = 8$, as they are both higher than $v_{2,1} = 7$, and agent 2 drops out of the market. At time t_2, the unit is again allocated to agent 1 (due to the marginal value of 3 being higher than 1). However, now the marginal payments vector required from agent 1 is $p_{-1} = \langle 0, 1, 7 \rangle$, while the marginal valuations are $v_1 = \langle 10, 8, 3 \rangle$. Given the prices, agent 1 prefers 2 units to 3 (because $10 + 8 - 1 > 10 + 8 + 3 - 1 - 7$), so the third is burnt. The overall market efficiency is lower, as the third available unit is now burnt, whereas with $r_1 = 1$ it was allocated to agent 2. Note, however, that even though the efficiency is much lower, agent 1 has an incentive to declare its true maximum charging speed $r_1 = 2$ as, in this case, its payment does not change.

4 Truthfulness Properties

In this section, we discuss the incentive compatibility properties of our mechanism under the following assumption:

Limited Misreports: *Agents cannot report an earlier arrival, a later departure, or a higher charging rate, i.e.,* $\hat{a}_i \geq a_i, \hat{d}_i \leq d_i, \hat{r}_i \leq r_i$ *must hold.*

In our domain, this is a reasonable assumption since a vehicle owner cannot physically plug in her vehicle before it is available or unplug it after the actual departure. Note that the assumption $\hat{r}_i \leq r_i$ is natural for EV charging. While most electric batteries can be configured to charge at a slower rate, charging them at a faster rate than the one allowed by the manufacturer might destroy them. As the battery is an integral and expensive part of an EV, this by itself acts as a natural deterrent. An EV agent may strategise by reporting a $\hat{r}_i$ lower than its true maximal speed, but we show truthful reporting of r_i is a dominant strategy. Given this, we can state the following:

Theorem 1. *Assuming limited misreports, and given the decision and payment policy as described above, reporting* $\hat{\theta}_i = \theta_i$ *is a dominant strategy for* $\forall i \in I$.

Proof Sketch: The proof for this theorem has 3 parts:

1. Showing that, after fixing any tuple of $\langle \hat{a}_i, \hat{d}_i, \hat{r}_i \rangle$, reporting the valuation vector $\hat{\mathbf{v}}_i = \mathbf{v}_i$ truthfully is dominant strategy incentive compatible (DSIC).

2. Showing that for any fixed $\langle \hat{\mathbf{v}}_i, \hat{r}_i \rangle$ (and under the limited misreports assumption $\hat{a}_i \geq a_i$, $\hat{d}_i \leq d_i$), reporting truthfully $\hat{a}_i = a_i, \hat{d}_i = d_i$ is DSIC.

3. Showing that for any fixed $\langle \hat{\mathbf{v}}_i, \hat{a}_i, \hat{d}_i \rangle$, using the limited misreports assumption $\hat{r}_i \leq r_i$, reporting truthfully the maximum charging rate $\hat{r}_i = r_i$ is DSIC.

The first two parts can be shown by using the same proof techniques as in [4]. For the third part, we can show that the vector of marginal payments $\mathbf{p}^{\langle \hat{r}_i \rangle}$ (computed given the report $\hat{r}_i$) will always contain a *subset* of the elements from the vector $\mathbf{p}_{-i}^{\langle r_i \rangle}$ (computed given the truthful report r_i), because $\hat{r}_i \leq r_i$. Thus, either an agent gets the same allocation, but pays less by reporting a higher speed (because the p_i vectors are increasingly ordered, and the payment is the sum of the first k_i units allocated), or the agent is allocated more units but then those additional units are needed, given the prices (otherwise, burning would apply).

5 Theoretical Bounds on Allocative Efficiency

An important question given the online allocation with burning proposed above is how the allocative efficiency compares to that of an optimal offline allocation (i.e., assuming full knowledge of the future). To answer this, we consider the cases with immediate burning and on-departure burning separately. For the case of immediate burning, we can show that no lower bound can be established through an example:

Example 1: Let an agent A_{syn} with marginal valuation vector $\mathbf{v}_{\text{syn}} = \langle v_1, v_2, \ldots, v_n \rangle$, where marginal valuations are strictly decreasing, i.e., $v_1 > v_2 > \ldots > v_n$. Assume supply is $S(t) = 1$ for all t, and at each time step, one other "local" agent is present for that time step only and desires a single unit. The valuation of the first local bidder is $\mathbf{v}_1 = \langle v_1 - \epsilon \rangle$, the valuations of the next two are $\mathbf{v}_{2,3} = \langle v_2 - \epsilon \rangle$, the next three $\mathbf{v}_{3,4,5} = \langle v_3 - \epsilon \rangle$, and so on, where ϵ is an arbitrarily small quantity. In other words, each marginal value $v_k - \epsilon$ appears exactly k times.

In this example, agent A_{syn} is pre-allocated every unit, but for each valuation v_k, the first $k - 1$ pre-allocated units are burnt, with only the last unit being allocated (due to the expanding $\mathbf{p}_{-i}^{\langle t \rangle}$ vector). Thus, in the limit, the fraction of units burnt goes to 1, while the efficiency goes to 0.

For on-departure burning, Theorem 2 provides the following worst-case guarantee:

Theorem 2. *The mechanism with on-departure burning is 2-competitive with the optimal offline allocation, for a setting with non-increasing marginal values.*

Proof Sketch: In order to establish a competitive bound with the optimal offline allocation, we use a "charging argument"' similar to Hajiaghayi et al. [5].[3] The basic idea is to "charge" (or match) all the marginal values of each agent that are allocated in the offline case with one or more values allocated online. Specifically, let $v_{i,k}^{\text{off}}$ denote the k-th marginal value of agent i that is allocated in the offline case. In the following we say that a unit is satisfied (or allocated) in the online case if it is actually allocated to agent i by the greedy mechanism and not burnt on departure.

Consider each marginal value $v_{i,k}^{\text{off}}$ from agents i that are allocated in the *offline* case, and to "charge" the values as follows: 1. Marginal values allocated both in the online and offline case are charged to themselves. 2. Marginal values that are allocated in the offline case, but not in the online case will always be displaced by some higher value $v_{j,p}^{\text{off}}$. Note that, since these values could be allocated online and are higher, they must necessarily also be allocated in the offline case (but this may occur at a different time than in the online case, when they would displace a lower valued unit). To complete the charging argument, we need to show what happens to these units w.r.t. burning.

Lemma 1. A marginal utility value $v_{i,k}^{\text{off}}$ that is allocated to agent i *both* offline and online cannot be burnt on the departure of agent i in the online market. Moreover, if a unit $v_{i,k}^{\text{off}}$ is displaced by another unit $v_{j,p}^{\text{off}}$, the displacing unit $v_{j,p}^{\text{off}}$ cannot be burnt on departure of agent j.

Proof Sketch: The proof for this lemma relies on comparing the vectors of decreasing marginal values $\mathbf{v}_i$ and the payment vector $\mathbf{p}_{-i}$ as well as the condition for burning specified by the on-departure burning decision policy. In summary, one can show by contradiction that, if a unit $v_{i,k}^{\text{off}}$ is allocated both offline and online, there could not be

[3] Here the term "charging" does not refer to electricity charging, but represents the name of a proof device used in online MD.

enough higher valued marginal units in agent i's active interval $[a_i, d_i]$ for this unit (or a higher valued one) to be burnt.

Given Lemma 1, for all agents i, each marginal value unit that is allocated offline can be charged *at most* twice:

- At most once to itself (if it is also allocated online).

- At most once by another unit allocated offline, with a value less than its own, that it displaced at some time t when that unit was allocated in the offline case (recall that in the mechanism with on-departure burning, each unit $v_{i,j}$ can be the winning bid at most once, so it can displace at most one other unit). Thus each unit allocated offline is charged at most twice, giving the online allocation with on-departure burning and non-increasing marginal values a competitive bound of 2 with the optimal offline allocation.

Theorem 2 shows the multi-unit demand case with on-departure burning is no worse in terms of worst-case competitive bound than the case with single unit demand (c.f. Hajiaghayi et al. [5]) which may seem surprising, given the burning. However, the only units allocated online which could be burnt online on departure of an agent i are units of such low value that would not be allocated in the corresponding offline case.

6 Experimental Evaluation

In this section, we apply our mechanism to a range of settings that are based on real data collected during the first large-scale trial of EVs in the UK. Our experiments examine two main questions. First, we study how our online allocation mechanism performs compared to a range of benchmarks as demand for electricity increases in a neighbourhood with limited supply. Second, we look at how the gradual introduction of fast-charging EVs would affect the neighbourhood, both in terms of social welfare (i.e., overall fuel savings) and the financial savings of individuals.

6.1 Experimental Setup

We base our experiments on data gathered by the CABLED (Coventry And Birmingham Low Emissions Demonstration) project,[4] which is the first large-scale endeavour in the UK to record and study the driving and charging behaviours of EV owners. The arrival and departure times of vehicle owners are generated based on this data. We also construct the agents' valuation vectors using the expected travel distances reported by the project, as well as the fuel and electric efficiencies and battery capacities of typical plug-in hybrid EVs, such as the Chevrolet Volt or Toyota Plug-In Prius. Such hybrid electric vehicles have a dual engine, and a per mile driving efficiency, which can be expressed either in units of electricity or litres of fuel (assuming a standard driving speed). As such, an agent's marginal valuation of a unit of electricity corresponds to the savings in fuel costs that the agent expects to make due to consuming this unit instead of regular fuel (based on a fuel price of £1.3/litre). For each configuration, 500 or 1000 runs were performed, with the expected driving distance for each vehicle for the next day being and its entry point into the market being randomized in each run. For full details, see Gerding et al. [4], where we use the same setup and dataset.

For each experiment, we consider a single day and, to calculate the capacity constraints (the supply function $S(t)$), we divide the day into 24 hourly time steps. For each time step, we first obtain the overall neighbourhood consumption based on real

[4] See `http://cabled.org.uk/`

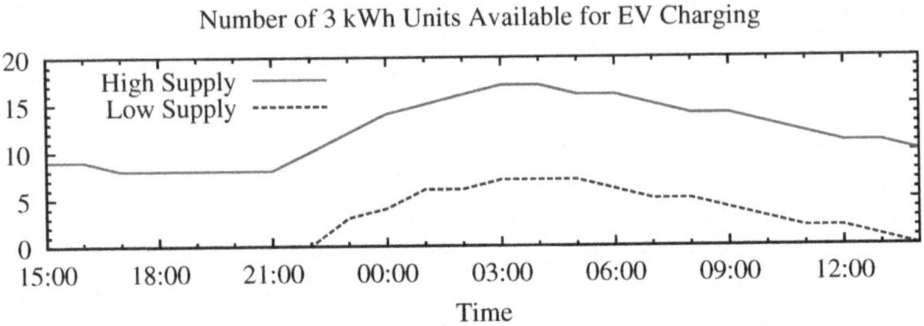

Fig. 2. Units of electricity available for EV charging

data.[5] We then consider two possible scenarios. First, when supply is *low*, electricity is highly constrained, and we set the capacity limit to 90% of the peak overall consumption, i.e., about 0.9 kW per household. This scenario represents neighbourhoods where the local transformer can only just support domestic demand with no spare capacity for vehicle charging during peak hours. Second, for *high* supply with significant spare capacity, we set the limit to 150% of peak consumption, i.e., approximately 1.5 kW per household.

In addition to our mechanism with *On-Departure* and *Immediate* burning, we consider a set of benchmark mechanisms.

(i): *Random* allocates units randomly to agents that have some positive marginal valuation for them.

(ii): *Fixed Price* allocates units at random to agents that have valuations greater than a fixed price and collects a payment equal to this fixed price. The price is chosen a priori to optimise the expected social welfare (i.e., total fuel savings), given full knowledge of the agent type distributions.

(iv): *Heuristic* is a common earliest-deadline-first scheduling heuristic that allocates units to agents to maximise a weighted sum of the agent's valuation for the unit and its deadline. The weights are chosen as for *Fixed Price*.

(v): *Optimal* allocates units to optimise the overall social welfare, assuming full knowledge of all future arrivals.

Note that apart from the *Random* baseline, these benchmarks assume additional knowledge either about agent type distributions or even future arrivals, and so they should be seen as upper bounds for our approach rather than actual realistic alternatives. Furthermore, only *Random* and *Fixed Price* are truthful (DSIC) like our mechanism.

6.2 Result 1: General Trends with Increasing Demand

First we consider the effect of rising electricity demand within a neighbourhood. To this end, Figure 3 shows the performance of our mechanism in a neighbourhood of 25 households with increasing numbers of EVs.[6] These results are for the low supply setting only, as the variation of EV numbers also changes the supply/demand balance. We assume that electricity is allocated in units of 3 kWh (the charging rate of a standard

[5] We use the average data for domestic households in June 2010, as reported by SCE (`http://www.sce.com/`)

[6] Note that as this exceeds 25, some households will have multiple cars. We show this to indicate the effect of very high demand and assume there is no collusion within a household.

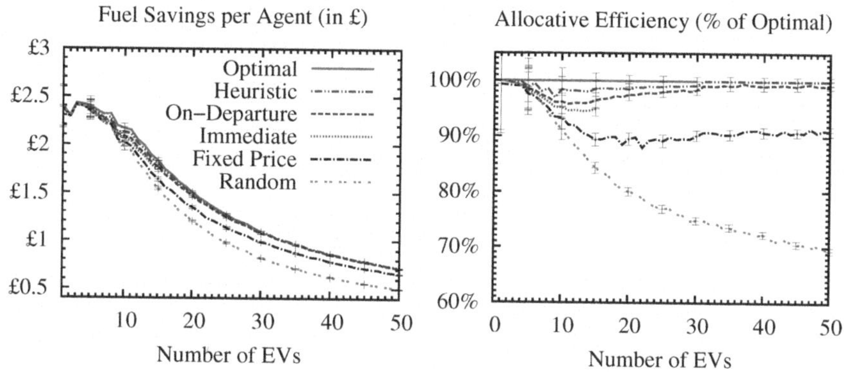

Fig. 3. Effect of increasing demand for electricity in a small 25-household neighbourhood with low supply

UK power socket), and that maximum rates are chosen at random from $\{1,2,3\}$ units, corresponding to the rates of currently available fast chargers for domestic use. Due to the computational complexity of the *Optimal* and *Immediate* mechanisms, we only plot those up to 30 and 15 agents, respectively.[7]

First, the graph on the left shows the daily fuel saving (excluding payments) for EV owners under the different mechanisms (all results reported are averaged over at least 500 samples and plotted with 95% confidence intervals). This shows that there is little difference between the various mechanisms when competition is low, as most agents are allocated; but as demand increases, our mechanism starts to outperform the other truthful benchmarks (by almost 50%) and achieves 95% or more of the optimal throughout. Interestingly, our mechanism sustains higher savings per agent for neighbourhoods with far higher EV penetration than the truthful benchmarks. For example, to guarantee a saving of at least £0.7 (just over half a litre of fuel), the *Random* benchmark can support up to 35 EVs, while our mechanism can cope with 50. The graph on the right shows the overall allocative efficiency achieved in the same setting, normalised to the optimal (beyond 30 EVs, this is normalised to the *Heuristic* as a close approximation).

Our principal finding is that online allocation with *On-Departure* burning achieves almost the same performance as the *Optimal*, calculated using full prior information about departures (the difference is not statistically significant). Moreover, the difference between our online allocation with *Immediate* burning and with *On-Departure* burning is also not significant. Surprising, given the different theoretical performance bounds of these policies, and indicates that burning is not a large problem in realistic settings.

6.3 Result 2: Proliferation of Fast-Charging EVs

We now consider in more detail how the presence of fast-charging vehicles affects market performance. To this end, we examine a larger neighbourhood of 50 households with one EV each. We choose this size, because it is realistic and leads to more interesting results, due to increasing likelihood of competition at each time step. Due to

[7] Briefly, *Immediate* is computationally hard, because the price vector $\mathbf{p}_{-i}^{\langle t \rangle}$ has to be computed at every time step by simulating the market without agent i, and, recursively, for every agent within that market, as it is needed to compute when burning takes place.

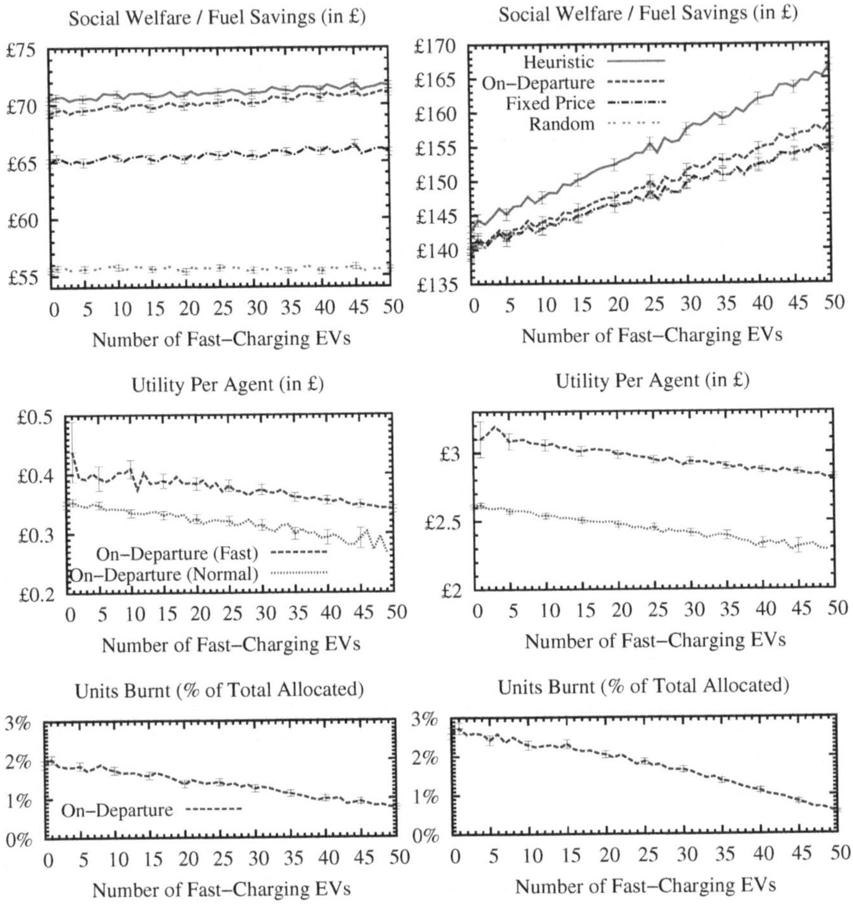

Fig. 4. Social welfare, individual utilities and units burnt in scenarios with low supply (left) and high supply (right), as more fast-charging EVs are introduced

their computational cost, we no longer run the *Optimal* and *Immediate* mechanisms (but their performance is similar to the *Heuristic* and *On-Departure*, respectively, as shown previously). To investigate the impact of fast-charging, we assume there are two agent types — the first, *normal*, can charge a single unit of 3 kWh per time step, while the second, *fast*, are equipped with fast chargers that can charge up to three such units per time step. Throughout the experiments, we vary the number of fast-charging EVs (out of the total 50).

Results for this setting are shown in Figure 4. First, we note that the trends for the two scenarios are different – when supply is low, the introduction of more fast charging vehicles results only in a small overall improvement in social welfare. However, when it is high, there is a very marked improvement, with an additional 13–14 litres of fuel per day being saved by our mechanism as more fast-chargers are present (corresponding to a saving of around 5000 litres of fuel per year over the whole neighbourhood). This is because, at low demand times, there is some spare capacity that remains unallocated, unless vehicles have a high charging speed.

With respect to the utility of individual agents (including payments to the mechanism), we see that agents *in both settings* always have an incentive to switch to fast-charging EVs (e.g., by purchasing a domestic fast charger). With low supply, the expected daily saving when switching to a fast-charging EV is approximately £0.05, while with high supply, this is around £0.45. In both cases, this benefit is the result of increasing available supply per time step, as well as increasing the size of the price vector.

Finally, another benefit of introducing fast-charging vehicles is the reduced proportion of units of electricity that are burnt by our mechanism. In the low supply setting, the percentage of burnt units is more than halved to less than 1% of all units allocated (on average 1.5 kWh) when all cars are fast-charging. With high supply, this is even more significant as only about a third of units are burnt. This reduction is a result of reduced competition and a larger price vector.

7 Conclusions and Further Work

This work makes several key contributions to the existing literature. On the theoretical side, we extend the multi-unit, online mechanism proposed in Gerding et al. [4] to also allow for modeling multi-unit demand *per time step*, in conjunction with multi-unit demand over time. For our EV application, this allows us to model more realistic markets, which include vehicles with heterogeneous charging capabilities. Moreover, we provide the first theoretical worst-case bounds for multi-unit online mechanisms. On the practical side, we build a detailed simulation of an EV charging market and show that our online mechanism performs close to the offline optimal in a variety of realistic settings. We also show that faster-charging batteries lead to savings in fuel consumption and increased allocative efficiency.

In future work, we plan to compare the model-free mechanism presented in this paper with online mechanisms that use a model of expected future arrivals. Moreover, we are interested in studying whether there are truthful mechanisms for this problem that do not require monetary payments.

Acknowledgements. This work was carried out as part of the IDEaS project (*http://www.ideasproject.info*), funded by a company active within the energy domain.

References

1. Clement, K., Haesen, E., Driesen, J.: Coordinated charging of multiple plug-in hybrid electric vehicles in residential distribution grids. In: Proc. of the IEEE/PES Power Systems Conference and Exposition, pp. 1–7 (2009)
2. Constantin, F., Feldman, J., Muthukrishnan, S., Pal, M.: An online mechanism for ad slot reservations with cancellations. In: Proc. SODA 2009, pp. 1265–1274 (2009)
3. Friedman, E., Parkes, D.C.: Pricing WiFi at Starbucks– Issues in online mechanism design. In: 4th ACM Conference on Electronic Commerce, pp. 240–241 (2003)
4. Gerding, E.H., Robu, V., Stein, S., Parkes, D.C., Rogers, A., Jennings, N.R.: Online mechanism design for electric vehicle charging. In: Proc. of the 10th Internetional Conference Autonomous Agents and Multi-Agent Systems, AAMAS 2011 (2011)

5. Hajiaghayi, M., Kleinberg, R., Mahdian, M., Parkes, D.C.: Online auctions with re-usable goods. In: 6th ACM Conference on Electronic Commerce (EC 2005), pp. 165–174 (2005)
6. Juda, A.I., Parkes, D.C.: An options-based solution to the sequential auction problem. Artificial Intelligence 173, 876–899 (2009)
7. Koslowski, T.: The electric vehicle's value chain and technology evolution. Technical report, Gartner Inc. (2009)
8. Parkes, D.C., Singh, S.: An MDP-Based approach to Online Mechanism Design. In: Proc. of NIPS 2003, pp. 791–798 (2003)
9. Porter, R.: Mechanism design for online real-time scheduling. In: Proc. 5th ACM Conference on Electronic Commerce (EC 2004), pp. 61–70 (2004)
10. Shao, S., Pipattanasomporn, M., Rahman, S.: Challenges of PHEV penetration to the residential distribution network. In: IEEE Energy Society General Meeting (2009)

Author Index

The manufacturer's authorised representative in the EU is Springer Nature Customer Service Centre GmbH, Europaplatz 3, 69115 Heidelberg, Germany. If you have any concerns regarding our products, please contact ProductSafety@springernature.com

Printed and bound by CPI Group (UK) Ltd, Croydon, CR0 4YY
25/09/2026
02222895-0005